Wireless Networks

Series Editor

Xuemin Sherman Shen
University of Waterloo, Waterloo, ON, Canada

The purpose of Springer's Wireless Networks book series is to establish the state of the art and set the course for future research and development in wireless communication networks. The scope of this series includes not only all aspects of wireless networks (including cellular networks, WiFi, sensor networks, and vehicular networks), but related areas such as cloud computing and big data. The series serves as a central source of references for wireless networks research and development. It aims to publish thorough and cohesive overviews on specific topics in wireless networks, as well as works that are larger in scope than survey articles and that contain more detailed background information. The series also provides coverage of advanced and timely topics worthy of monographs, contributed volumes, textbooks and handbooks.

** Indexing: Wireless Networks is indexed in EBSCO databases and DPLB **

More information about this series at http://www.springer.com/series/14180

Zhou Su • Yilong Hui • Tom H. Luan • Qiaorong Liu
Rui Xing

The Next Generation Vehicular Networks, Modeling, Algorithm, and Applications

Springer

Zhou Su
Shanghai University
Shanghai, China

Yilong Hui
Xidian University
Xi'an, Shaanxi, China

Tom H. Luan
Xidian University
Xi'an, Shaanxi, China

Qiaorong Liu
Shanghai University
Shanghai, China

Rui Xing
Shanghai University
Shanghai, China

ISSN 2366-1186   ISSN 2366-1445   (electronic)
Wireless Networks
ISBN 978-3-030-56829-0   ISBN 978-3-030-56827-6   (eBook)
https://doi.org/10.1007/978-3-030-56827-6

This Springer imprint is published by the registered company Springer Nature Switzerland AG
The registered company address is: Gewerbestrasse 11, 6330 Cham, Switzerland

Preface

The next generation vehicular networks are expected to provide services with highly improved network performance in terms of low cost and latency as well as high network efficiency and quality of experience (QoE) to vehicles. Towards better network maintainability and sustainability, this book proposes the novel network envisions and framework design principles to realize the goals of the next generation vehicular networks. As it is widely recognized, enabling technologies such as information centric networks (ICN), edge caching, computation offloading, artificial intelligence (AI), and autonomous driving play important roles in the wireless technology developments along with the research on intelligent transportation systems (ITS) and smart cities. The investigation and development on the integration of vehicular networks and the new enabling technologies thus provide important references for designing and developing the next generation vehicular networks.

However, with the advance of vehicular applications and the diverse service requirements of vehicles, to develop the next generation vehicular networks by investigating the integration of vehicular networks and the enabling technologies becomes a new challenge. The reasons that contribute to this are detailed as follows: (1) There is a huge amount of content to be delivered and vehicle users typically have different behaviors in the vehicular networks. With the integration of ICN and vehicular networks, an analytical scheme to facilitate the content delivery in information centric vehicular networks is thus needed, where the behaviors of vehicle users in the content delivery process should be taken into account. (2) With the massive demands of various kinds of content services and different social relationships among vehicles, the edge caching in vehicular networks faces the challenges to decrease the latency, increase the efficiency of the networks, and improve the QoE of vehicles. To this end, a novel edge caching framework by considering the service requirements and the social ties of vehicles is necessary to be developed. (3) In the vehicular networks, vehicles typically need to execute various computing tasks during driving, such as video processing, online social network services, and virtual reality. Caused by the limited computation power, vehicles need to offload a part of their tasks to edge devices. Since the emerging network applications may have different requirements of computing resources, the model of optimal computation resource allocation becomes a challenge to provide vehicles with satisfactory computing services. (4) In the vehicular networks, a vehicle which intends to offload its computing task may connect to more than one edge servers, where some edge servers may declare unreasonable prices to execute the computing task. In addition, the edge servers may maliciously declare a low price to complete the task with low quality. Therefore, a secure computation offloading scheme needs to be designed to constrain the bids of edge servers and guarantee the computing service with high quality. (5) With the popularization of the vehicular networks, malicious nodes may attack the vehicles and RSUs in the content delivery process. Therefore, secure content delivery to protect the vehicles against threats (i.e., outside attacks and inside attacks) and provide a secure content delivery environment becomes a challenge. (6) Due to the huge data perceived from complicated traffic environment and the limited computing power of vehicles, it is quite challenging to achieve vehicular networks enabled autonomous driving, which is one of the most attractive applications in the next generation vehicular networks. To this end, an autonomous driving scheme in which AVs can learn and drive with groups needs to be designed.

As an effort to address these issues, this book focuses on the key enabling technologies to design the framework for the next generation vehicular networks to satisfy various vehicular services requested by vehicles. The network architectures and the framework design principles are discussed in depth including the analysis of reputation based content delivery in information centric vehicular networks, contract based edge caching in vehicular networks, Stackelberg game based computation offloading in vehicular networks, auction game based secure computation offloading in vehicular networks, bargain game based secure content delivery in vehicular networks, and deep learning based autonomous driving in vehicular networks.

Specifically, in this book, we focus on the modeling, algorithms, and applications in the next generation vehicular networks, in order to improve the performance of vehicular services and facilitate the future smart city transportation system. In Chap. 1, we introduce the overview of vehicular networks and the enabling technologies that can be integrated with vehicular networks to pave the way for the integration of the vehicular networks and different technologies. In Chap. 2, with the analysis of the various behaviors of vehicles, we discuss the reputation based content delivery in information centric vehicular networks. In Chap. 3, by considering the social relationships among vehicles, we develop a contract based edge caching scheme in vehicular networks to enhance the content delivery performance. In Chap. 4, we study an incentive mechanism by designing the Stackelberg game model to analyze the computation offloading problem in vehicular networks. With this mechanism, both the vehicle and mobile edge computing (MEC) server are motivated to obtain the optimal game strategy. In Chap. 5, by integrating edge computing and cloud computing, we propose a secure computation offloading scheme to help the vehicle select the optimal edge server to offload its task. By using the proposed scheme, the bid prices of edge servers can be constrained and the service quality of the computing task can be guaranteed. In Chap. 6, we propose a security aware content delivery scheme in vehicular networks. Specifically, we first establish a trust evaluation scheme by introducing authority units (AUs) to monitor the actions of both vehicles and RSUs during the process of content delivery. Then, a price competitive scheme between vehicles and RSU is proposed by using the bargain game to encourage them to improve their trust values and utilities. In Chap. 7, we talk about the vehicular networks enabled autonomous driving and investigate the framework of deep learning for collaborative autonomous driving. Finally, in Chap. 8, we conclude this book and summarize some future research directions in the next generation vehicular networks.

This book validates the network architectures and the framework design principles for reputation based content delivery in information centric vehicular networks, contract based edge caching in vehicular networks, Stackelberg game based computation offloading in vehicular networks, auction game based secure computation offloading in vehicular networks, bargain game based secure content delivery in vehicular networks, and deep learning based collaborative autonomous driving in vehicular networks, where the performances of the above studies are evaluated by simulations. Therefore, this book can provide valuable insights on the integration of conventional vehicular networks and the enabling technologies in depth to satisfy the goals of the next generation vehicular networks.

We would like to thank Prof. Dongfeng Fang at the Department of Computer Science and Software Engineering, Cal Poly, San Luis Obispo, USA, for her valuable discussions and insightful comments. We would like to thank Dr. Yuntao Wang at the School of Cyber Science and Engineering of Xi'an Jiaotong University, Xi'an, China, for his constructive comments and helpful discussions. We would like to thank Prof. Qichao Xu, Dr. Minghui Dai, Dr. Weiwei Li, and Dr. Hui Zeng at the School of Mechatronic Engineering and Automation of Shanghai University, Shanghai, China, for their valuable discussions and helpful comments. We would also like to show our special thanks to the staff at Springer Science+Business Media: Ms. Susan Lagerstrom-Fife and Ms. Jennifer Malat, for their kind help throughout the publication and preparation processes.

Shanghai, China   Zhou Su

Xi'an, China   Yilong Hui

Xi'an, China   Tom H. Luan

Shanghai, China   Qiaorong Liu

Shanghai, China   Rui Xing

Contents

Acronyms

5G	5th generation mobile networks
AI	Artificial intelligence
AUs	Authority units
AVs	Autonomous vehicles
BS	Base station
CCNs	Content centric networks
CPS	Cyber physical system
CPU	Central processor unit
CS	Content store
DPoS	Delegated proof of stake
DSRC	Dedicated short range communication
EMS	Element management system
ETSI	European Telecommunications Standards Institute
FBSs	Femto base stations
FIB	Forwarding information base
GPS	Global positioning system
ICNs	Information centric networks
IoTs	Internet of Things
ITS	Intelligent transportation system
LCU	Logic control unit
LTE	Long term evolution
MBSs	Micro base stations
MCC	Mobile cloud computing
MEC	Mobile edge computing
NBI	Northbound interface
NFV MANO	NFV management and orchestrator
NFV	Network function virtualization
NFVI	NFV infrastructure
NHTSA	National Highway Traffic Safety Administration
OBU	On board unit
PBFT	Practical Byzantine fault tolerance

PIT	Pending interest table
PoS	Proof of stake
PoW	Proof of work
QoE	Quality of experience
RSU	Roadside unit
SBI	Southbound interface
SDN	Software defined network
V2I	Vehicle to infrastructure
V2V	Vehicle to vehicle
VMs	Virtual machines
VNFs	Virtual network functions
VSNs	Vehicular social networks

Chapter 1
Introduction

1.1 Overview of Vehicular Networks

1.1.1 Architecture of Vehicular Networks

As one of the most important enabling technologies to realize the next generation intelligent transportation system (ITS), vehicular networks are considered as a set of vehicles embedded with on-board units (OBUs) and road infrastructures (i.e., roadside units (RSUs) and base stations (BSs)) [1–5]. With a radio interface, each OBU can make the connection with other OBUs, RSUs, BSs and other smart devices, by which they can communicate with each other to share useful information with the goal of facilitating the driving and transportation system. Generally, as the typical scenario which is shown in Fig. 1.1, vehicular networks mainly consist of two types: (1) vehicle to vehicle communications, i.e., V2V, and (2) vehicle to roadside infrastructure communications, i.e., V2I [6–10].

For V2V communications, a vehicle can communicate with the vehicles within its communication coverage directly via a single hop connection [11–15]. In addition, a group can be formed by vehicles to achieve the long distance communication via opportunistic routing through a multiple hop connection. In this way, a vehicle can connect with other vehicles which are outside of its communication coverage. For V2I communications, both the RSUs and BSs have the communication coverage so that they can communicate with the vehicles when the vehicles drive in the coverage [16–20]. Vehicular networks integrate the V2V and V2I by which the information can be smoothly shared in the networks. For example, if a vehicle is suffering a traffic accident, with the help of V2V communications, all vehicles connected with it can obtain this accident information. In addition, the accident information can be further relayed to other vehicles and RSUs or BSs to re-plan the driving routes. With the real time information sharing, the traffic accidents can be reduced and the traffic efficiency can be improved.

© Springer Nature Switzerland AG 2021
Z. Su et al., *The Next Generation Vehicular Networks, Modeling, Algorithm, and Applications*, Wireless Networks, https://doi.org/10.1007/978-3-030-56827-6_1

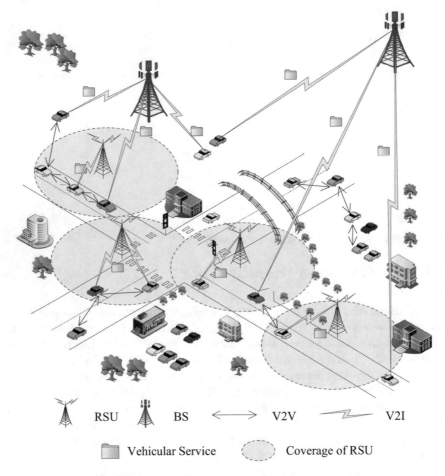

Fig. 1.1 An illustration of vehicular networks

Next, we introduce the two typical kinds of communication modes in vehicular networks in detail.

- **V2V communications:** This communication paradigm enables vehicles to connect and exchange information with each other directly without sending data to RSUs, BSs or the core networks [21–25]. Due to the opportunistic links among vehicles, V2V communications are considered as an important technology especially when the RSUs or BSs are not available. Besides, V2V communications are more economical than V2I communications in terms of the construction cost. With V2V, vehicles can not only share information with small data size such as accident warnings and lane change warnings, but also can share content with large data size such as popular movie trailer and video stream with the adoption of carry-and-forward strategy [26–28]. Specifically, when a vehicle moves to the communication coverage areas of other vehicles, the information and contents can be exchanged. Otherwise, the vehicle carries the information and waits for the next connection [29].

- **V2I communications:** This type of communication mainly provides a connection between OBUs placed at vehicles and infrastructures located along the roads. The infrastructures typically refer to the cellular networks (4G/5G and long term evolution (LTE)) e.g., BSs, and WiFi-like access units e.g., RSUs [30–33]. Both the RSUs and BSs are connected to the remote server using wired links. Compared with the RSUs, the BSs have larger communication coverage. In addition, the BSs can provide vehicles with a stable download rate while the RSUs only provide occasional connections due to the chaos contentions among vehicles. Through the V2I connection, vehicles are allowed to request various vehicular services by connecting to the RSUs or BSs.

With the integration of V2V and V2I, on one hand, notifications such as traffic and weather conditions or some specific road messages (e.g., maximum speed limit and overtaking warning) can be broadcasted to inform the nearby vehicles in vehicular networks to facilitate the transportation system. On the other hand, vehicular networks have the ability to provide entertainment services containing huge data for vehicle users to enhance their quality of experience (QoE) [34–38]. In what follows, we will detail the applications that can be provided with the deployment of vehicular networks.

1.1.2 Applications in Vehicular Networks

The applications supported in vehicular networks mainly include vehicle safety applications (e.g., automatic collision warning, remote vehicle diagnostics, emergency treatment and assistance for safety driving), vehicle traffic applications (e.g., event notification, traffic scheduling and path planning), and infotainment applications (e.g., high-speed Internet access and multimedia content sharing) [39].

- **Safety applications:** Safety applications are supported by the vehicular networks to reduce the risk of road accidents and the number of the people injured and killed on the roads [40]. According to the statistic information, there are approximately 2.5 million accidents every year and the second largest category of accidents is caused by rear end collisions [41]. By providing related road and vehicle information and assistance to drivers, road safety applications can help vehicles avoid accidents and collisions. Safety applications contain many aspects such as intersection collision warning, head on collision warning and rear end collision warning. All of these warnings can be detected by vehicles or infrastructures and the information can be further shared in the networks. In addition to collision warnings, safety applications can also support other warnings and assistances, such as overtaking warning, emergency vehicle warning, pre-crash sensing, wrong way driving warning, signal violating warning and lane change assistance [42]. For example, a vehicle can integrate the information from other vehicles in the same lane to avoid collisions. If the vehicle intends to change its driving lane, by using the messages from the adjacent vehicles in the neighboring lane, it can complete the lane change in safety.

- **Traffic applications:** This type of application aims at improving the driving experience by optimizing the traffic flow and providing vehicle position information and maps [43]. Compared with safety applications, traffic applications have no strict latency and reliability requirements. These applications may need to be supported by a large number of alert messages. For instance, most of the accidents are caused by the over speed of vehicles. Therefore, the vehicle speed management becomes particularly important. In this case, the alert messages (e.g., the regular speed limit notification) can help vehicle drivers pay attention to their driving speeds. In addition, this type of application can greatly enhance the traffic efficiency. One of the cases is that the cooperation detection and navigation of vehicles and roadside infrastructures can help a driver obtain the optimal path to the destination by collecting the traffic information in the vehicular networks. In addition, by regulating and scheduling traffic flow, the traffic efficiency can be significantly improved.

- **Infotainment applications:** The applications belong to this category is to provide on-demand entertainment services to the driving vehicles, which includes custom information services, Internet access, multimedia content sharing, video streaming, and so on. In recent years, video contents and video applications have become the increasing infotainment services, such as remote video conferences, live news and films. In fact, with the deployment of vehicular networks, vehicles are no longer a simple means of transportation for people, but a private place to support entertainment services. In this way, the time of attendance can be fully used to improve the driving experience.

1.2 Overview of Enabling Technologies

1.2.1 Advanced Communication-5G

The fifth generation (5G) mobile communication technology is the latest generation of cellular mobile communication technology. As the next telecommunication standard, the 5G wireless system has attracted more and more attention from academia, industry and government [44–47]. The development of 5G comes from the growing demand for mobile data. Specifically, with the development of mobile Internet, the networks exhibit two notable features. One feature is the growing number of connected devices. Another feature is the emergence of new services and applications requested by the connected devices. Meanwhile, the skyrocketing mobile data traffic poses serious challenges to the network:

- According to the current development of mobile communication networks, on one hand, it is difficult to support the growth of data traffic. On the other hand, the network energy consumption is unable to efficiently support the new services and applications.
- The traffic growth will inevitably lead to further demand for the spectrum. However, the available mobile communication spectrum is scarce. In addition, the fragmented distribution makes it difficult to achieve efficient use of the spectrum.
- The future network is a heterogeneous mobile network. To improve network capacity, it is necessary to solve the problem of integrating heterogeneous networks, simplifying interoperability and enhancing user experience.

The 5G mobile communication networks, which are composed of several heterogeneous base stations, such as micro base stations (MBSs) and femto base stations (FBSs), are expected to solve the above challenges and meet the growing demand for mobile traffic. Compared with 4G mobile communication technology, 5G has the following new features:

- The peak data transmission rate reaches Gbit/s which is faster than the current wired Internet and 100 times faster than the previous 4G cellular network.
- By taking various Internet of things (IoTs) devices into account, 5G has a large network capacity, which can provide the connection capacity of 100 billion devices to support IoT communication.
- With the heterogeneous base stations, 5G can provide a wide-area communication coverage, where the user experience rate in 5G is expected to reach 100 Mbit/s.

Based on the features of 5G, we can know that the 5G networks can solve the above challenges efficiently. In addition, with a wide coverage of networks, the information sharing between vehicles and infrastructures becomes easier and faster. Consequently, the efficiency of data transmission in the vehicular networks can be significantly improved. Furthermore, with the support of high-speed data transmission, some new applications in the vehicular networks can be realized and deployed, such as on-line gaming, virtual reality and autonomous driving.

1.2.2 Mobile Edge Computing

Currently, mobile edge computing (MEC) technology is emerged as a new paradigm to overcome the challenges brought by conventional mobile cloud computing (MCC) technology [48–52]. The concept of MEC was firstly proposed by the European Telecommunications Standard Institute (ETSI) in 2014 and it can be deployed at RSUs and BSs to provide computing services in close proximity to mobile subscribers. Due to the physical proximity, a large number of vehicles can offload their computation tasks to the MEC servers instead of accessing the remote

Table 1.1 Comparison of
MCC and MEC

Characteristics	MCC	MEC
Deployment	Network core	Network edge
Computing	Remote	Local
Service support	Virtualization	Virtualization
Architecture	Centralized	Distributed
Mobility	No/Medium	Yes
Location awareness	No/Medium	Yes
Scalability	Medium	High
Availability	High	High
Construction cost	High	Low

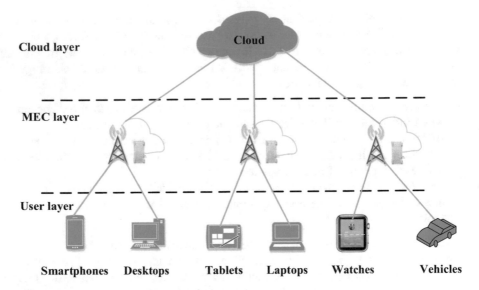

Fig. 1.2 Typical three-layer service model of MEC

cloud. In this way, the traffic congestion on the backbone networks thus can be
alleviated [53–55]. Furthermore, the latency of the computation tasks can be greatly
reduced compared with that executed by the remote cloud servers so that the QoE
of the vehicles can be guaranteed. Besides, with the distributed deployment of the
MEC servers, a large number of computation demands generated by vehicles can
be responded simultaneously [56], where the efficiency of the networks can be
improved. The detailed comparison of MCC and MEC is shown in Table 1.1.

A three-layer service model of MEC is depicted in Fig. 1.2. From this figure, we
can see that the framework consists of three layers, which are user layer, i.e., various
types of mobile devices (e.g., smart phones, desktops, laptops, tablets, watches
and vehicles), MEC layer and cloud layer. In vehicular networks, vehicle users
typically have different task computing demands. Due to the physical proximity
between MEC servers and vehicles, each task computing service can be responded

and supported by MEC server immediately. After the execution of the task has been accomplished, the MEC server then returns the result to the vehicle. In this way, the energy consumption related to radio access and dissemination latency can be greatly decreased compared with relying on the cloud server. However, with the low computing and executing power of MEC servers, they can not always meet all of the service demands generated from vehicles. In this case, the MEC servers can relay these computing demands to the cloud server. After the task is completed, the cloud server delivers the result of the computing task to the MEC server which will further deliver the result to the vehicle. In this way, for the urgent computing tasks, they can be executed by the MEC servers to reduce the latency time. For the tasks without the strict delay requirements, they can be completed by the cloud server to improve the resource utilization and task completion efficiency.

1.2.3 Network Function Virtualization

The network services are generally running on proprietary and dedicated hardware. Network function virtualization (NFV) [57–61] is a network architecture that aims to virtualize various types of network services. By consolidating different network equipment on standard high-volume servers and leveraging virtualization technologies, NFV has emerged as an initiative to transform the way of network functions in software [62]. In particular, it can easily be migrated from one network equipment to another without the need of having custom hardware appliances for each network function. In this way, the flexibility of network service deployment can be improved. In addition, compared with hardware installing, the investment cost of the network provider can be reduced [63]. For example, a virtual session boundary controller can be deployed to protect the networks without the construction cost and complexity for installing a physical network protection unit.

The NFV architectural framework [64], as shown in Fig. 1.3, generally consists of three parts which are virtual network functions (VNFs), NFV infrastructure (NFVI) and NFV management and orchestrator (NFV MANO), respectively.

- VNFs are software implementation versions of network function components running on multiple virtual machines (VMs). Each VNF is controlled by an element management system (EMS).
- NFVI provides the network environment, where VNFs are deployed and executed [65]. The virtual resources on NFVI are abstracted and logically partitioned from the underlying hardware resources through a virtualization layer, where the hardware resources can be computing, storage and network resources.
- NFV MANO provides organization and management of VNFs, which is controlled by a set of VNF managers. The physical and virtual resources of NFVI are also controlled through a virtualized infrastructure manager. In addition, NFV orchestrator is in charge of VNF and other network services so that they can achieve reasonable orchestration.

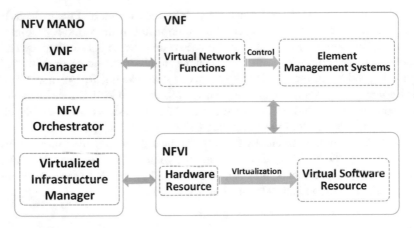

Fig. 1.3 NFV architectural framework

According to the analysis of the features of NFV, we then summarize the advantages of NFV-based vehicular networks as follows [66].

- With the adoption of NFV, the services in vehicular networks are virtualized as software and separated from hardware. As a result, the function update can be completed by only updating the software so that the updating process becomes easy and efficient.
- VNFs can be deployed dynamically and can support different networking types to provide services based on different network situations, which leads to a flexible deployment of network servers.
- With the NFV-based vehicular networks, the third parties also can join in the networks to develop and design new services, where the innovation of the service system will be accelerated. Furthermore, the reuse of virtualized functions in the system can reduce the time and investment of developing new services and applications.

1.2.4   Software Defined Network

Traditional networks are based on hardware and possess distributed or decentralized architecture [67–69], where the forwarding devices in the networks, e.g., routers, switches, etc., are used to forward the data or packets from the source to the destination. Software defined network (SDN) [70–74] is an emerging technology to provide high flexibility and scalability of networks, where the network is recognized as an operating system and the resources are managed by the SDN controller. On one hand, due to the separation of control plane and data plane, SDN can greatly improve the efficiency of vehicular networks and easily obtain logical centralization

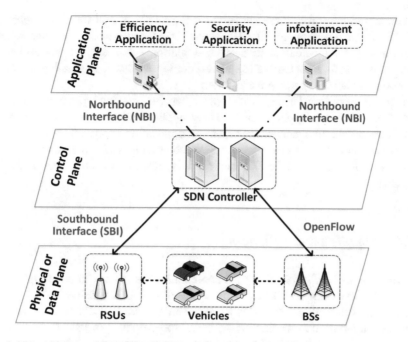

Fig. 1.4 Architecture of SDN-based vehicular networks

of network control [75]. In the SDN based vehicular networks, the RSUs and BSs are only in charge of providing services for vehicles. The control decisions are made by the SDN controller which has the global view of the networks. On the other hand, the programmability of the control makes it easier to create new abstractions in vehicular networks, simplify network management and promote network evolution [76–79]. Particularly, in the case of the SDN-based vehicular networks, the vehicles and the road infrastructures in the data plane are served as the forwarding devices [69].

As shown in Fig. 1.4, the network space can be decoupled into three planes, which are the data plane, control plane and application plane, respectively [80–83].

- **Data plane:** The data plane abstracts the underlying resources (e.g., vehicles) as SDN switches. These switches conform to the unified scheduling, follow the OpenFlow protocol and route traffic along the path towards the selected destination network [80]. In other words, the data plane does not concern about the control policy. It only needs to address the incoming and outgoing data packets according to the control instruction published by the control plane.
- **Control plane:** The control instruction is forwarded from each switch to the centralized controller, i.e., SDN controller. It executes direct control for the data plane elements through the well-defined southbound interface (SBI) [83]. The separation of the data plane and the control plane also benefits from the well-

defined programming interface between the switches and the SDN controller [71]. In the development of SBI technology, OpenFlow is the most notable example [84]. Its goal is to provide a platform that enables researchers to run experiments in networks [82]. Through OpenFlow, it is possible to control multiple switches by a single controller.

- **Application plane:** The application plane consists of many types of applications, such as efficiency applications, security applications and infotainment applications in vehicular networks [85]. Each application represents a client entity and may request services from the SDN controller, where the communication between the clients and the SDN controller is routed by the northbound interface (NBI).

1.2.5   Computation Offloading

With the help of MEC technology, the computation resources can be brought to the edge of the networks. In MEC, computation offloading is considered as one type of consumer-oriented service. This is because the vehicle users can benefit from MEC by computation offloading [86–89], e.g., speed up the process of computation, save computation resources, alleviate energy consumption and prolong battery life. For the integrated networks, the computation offloading aims to migrate some computation tasks or functions from vehicles to MEC servers. The computation offloading process contains three parts, which are offloading decision, offloading execution and results return, respectively.

In the vehicular networks, each vehicle typically has computing resources to execute its task by itself. In fact, vehicles on the road can form a group, which is called as vehicular cloud, to share their computing resources [90–93]. The resources owned by vehicles are however less than those of MEC server. For a vehicle which intends to compute its task, the first step is to make the task offloading decision. The decision indicates that the vehicle decides whether to offload its computing task to one of the MEC servers in the networks or not. Then, if the vehicle decides to offload its task, the task will be executed by its connected MEC server. After the task execution has been accomplished, the task result will be returned to the vehicle using the V2I communications.

Apart from the computation offloading decision, the vehicle also needs to determine the number of computation tasks that can be offloaded to its connected MEC server. Basically, a decision on computation offloading can result in two cases. One case is binary offloading, where the vehicle will offload all computation tasks or offload nothing. Another decision is to select a part of the computation tasks to offload, while the rest is processed by the vehicle itself. In addition, a task can be divided into several mutually independent subtasks or submodules, by which a part of the subtasks can be offloaded to the MEC server to speed up the task computing process. On the other hand, some subtasks can not be executed in a parallel way. For

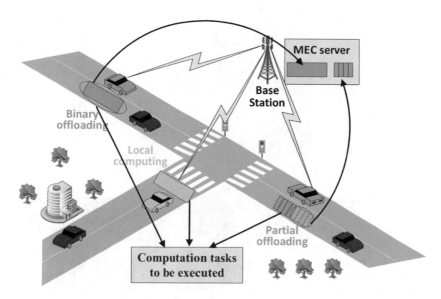

Fig. 1.5 Different computation offloading decisions

example, the input of task B may be based on the output of task A. If a user intends to execute task B, it should obtain the result of task A at first.

To summarize, the offloading decisions can be divided into three categories which are local computing, binary offloading and partial offloading, respectively. The details of different computation offloading decisions are shown in Fig. 1.5.

- **Local computing:** In the integrated networks, both end-user devices (e.g., smart phones, laptops and tablets) and IoT devices (e.g., vehicles equipped with OBUs) have computation capacities. Thus, the computation tasks of vehicles can be executed locally based on their own computation resources when the MEC servers are not available or the vehicles have enough time and computation resources to execute the computation tasks. Furthermore, if a vehicle does not want to give payment for executing offloading services, it can select local computing to execute the task.
- **Binary offloading:** The binary offloading means that all computation tasks cannot be partitioned and are offloaded and executed as a whole at the MEC server, where each computation task is required to be executed within the deadline. If the computation resources of the MEC server are not available, it can deliver the computation tasks to the cloud server for computing. After the offloaded tasks are completed, the results will be returned to the corresponding vehicles. The flow-process diagram of computation offloading is depicted in Fig. 1.6.

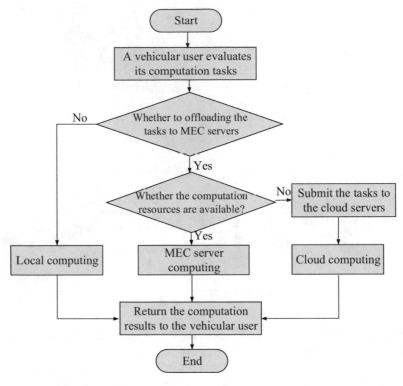

Fig. 1.6 The procedure of computation offloading

- **Partial offloading:** The computation tasks are executed by the vehicle and the MEC server together. In other words, the vehicle can execute a part of the tasks locally based on its own computation resources. The rest will be offloaded to one or more MEC servers for executing, as shown in Fig. 1.5. In practice, with the limited computation capacities, partial computation offloading is more flexible than the others so that many computation applications can be implemented with the adoption of partial computation offloading in vehicular networks.

1.2.6 Blockchain

Blockchain refers to a chain of blocks, which is essentially a decentralized database [94–98]. Specifically, as the underlying technology of Bitcoin, it is a series of data blocks generated by cryptography. The information of the sub-bitcoin network transaction is used to verify the validity of its information (anti-counterfeiting) and generate the next block.

Blockchain primarily addresses the trust and security of transactions, which proposes four technological innovations for this problem:

(1) Distributed ledger

In distributed ledgers, transaction accounting is done by multiple nodes distributed in different places, and each node records the complete accounts [99–101]. In this way, they can participate in supervising the legality of transactions, and can also jointly testify to the ledgers. Compared with the traditional distributed storage, the uniqueness of the distributed storage of the blockchain is mainly reflected in two aspects:

- Different from the traditional distributed storage, which always divides the data into multiple copies according to certain rules, each node of the blockchain stores the complete data according to the blockchain structure.
- The storage of each node in the blockchain is independent and equal in status. It relies on the consensus mechanism to ensure the consistency of storage.

In the networks, any node can not record the ledger data separately, which avoids the possibility that a single node who keeps the ledger will be controlled or be bribed to record a false ledger. In addition, the ledger is kept by all the nodes in the networks, thus ensuring the security of the ledger data.

(2) Asymmetric encryption and authorization technology

The transaction information stored in the blockchain is public, but the account identity information is highly encrypted and can only be accessed if authorized by the data owner, thereby ensuring data security and personal privacy. The process of asymmetric encryption is shown in Fig. 1.7. With the adoption of asymmetric encryption, information can be shared securely in the networks while protecting the privacy of vehicle users.

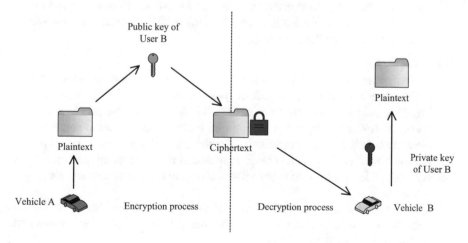

Fig. 1.7 The process of asymmetric encryption

(3) Consensus process

The consensus process is to reach a consensus among all the nodes who keep the ledger to determine the validity of a new record. This is both a means of identification and a means of tamper-proofing. There mainly exist four different consensus algorithms in blockchain that are applied to different application scenarios and strike a balance between efficiency and security:

- **Proof of work (PoW):** PoW requires the workstation to perform some time-consuming and complex calculations, and the answer can be quickly checked by the service provider. By using the time, equipment and energy as the guarantee cost, it can be ensured that the service and resources are used by real needs.
- **Proof of stake (PoS):** PoS is also known as equity proof. Similar to the property stored in bank, this model will assign a user the corresponding interest based on the amount and time of the digital currency the user owned.
- **Delegated proof of stake (DPoS):** The DPoS algorithm will elect several nodes from thousands of PoS nodes through a specific mechanism (such as the number of tokens they hold). Then, vote among them to elect the consensus nodes without choosing all nodes in the network to participate in the consensus process. This algorithm guarantees the efficiency of the consensus process while the time and energy consumption can be reduced.
- **Practical byzantine fault tolerance (PBFT):** PBFT is a state machine replica replication algorithm, in which the service is modeled as a state machine, and the state machine performs replica replication at different nodes of the distributed system. It is designed to allow the honest nodes in the system to cover the behavior of malicious nodes or invalid nodes. The number of nodes in the PBFT algorithm is fixed, one node represents one vote, and the Byzantine fault-tolerant calculation is implemented in a minority-substantial manner. The maximum fault tolerance is no more than 1/3 of the total number of nodes, which means that if there are more than 2/3 normal nodes, the whole system then can operate normally.

(4) Smart contract

The smart contract is a computer protocol designed to disseminate, verify and execute a contract in an informational manner [102–104]. It allows for trusted transactions without third parties, which are traceable and irreversible. Based on the non-tamperable data, smart contracts can execute some predefined rules and terms automatically. In the case of insurance, if everyone's information (including medical information and risk-generating information) is authentic, it is easy to operate automated claims in some standardized insurance products.

Recently, the use of blockchain technology in different scenarios in the vehicular networks has received extensive attention [105–110]. The reasons that contribute to this are as follows.

- Unlike the conventional mobile networks, vehicles in the vehicular networks have a high mobility, where vehicles often use opportunistic connections for networking. Consequently, vehicles in the vehicular networks are usually strange to each other.
- Since the resources (e.g., caching and computing) in the vehicular networks are not free so that many services are completed in the form of transactions. With the rapid development of the vehicular networks, the transactions are becoming more frequent than before.
- Vehicle users in the vehicular networks typically have various types. Being faced with different situations, they may have different behaviors which have a serious effect on the performance of vehicular services. For example, the aggressive vehicle users may broadcast the false information and the malicious vehicle users may steal the information from other vehicles.

Therefore, blockchain technology is needed for vehicular networks to ensure the security of information sharing and transactions.

1.2.7 Information Centric Networks

The current Internet is a point-to-point connection architecture based on packet switching between terminals. With the increasing demand for real-time video, voice and other communication services, due to the inherent structure of the TCP/IP system, the ability of the Internet to transmit and process this real-time multimedia data is increasingly challenged, resulting in low network efficiency and poor user experience. To address this issue, Information-centric Networking (ICN) has emerged and is expected to change the current end-to-end communication mechanism [111–115].

ICN breaks the host-centric connection mode of TCP/IP and becomes a mode of information (or content) centric. With ICN, the data will be independent of the physical location, and any node in the ICN network can be a content producer to generate contents. The existing studies related to ICN have some common goals: promote content distribution efficiency, improve network security, enhance large-scale network scalability and simplify the creation of distributed applications.

Content centric networking (CCN) is a specific implementation which is most studied in ICN. CCN is a receiver driven communication protocol that uses content as the core [116–121]. There are two kinds of packets in CCN, i.e., the content packet and the interest packet. The CCN node undertakes the storage, forwarding and routing tasks of the packets. A typical CCN node mainly includes a content store (CS), a pending interest table (PIT) and a forwarding information base (FIB).

- **CS:** The CS is similar to the IP router's cache, but the cached content will not be emptied after each communication. In this way, the content can be used for the next communication. This is a critical concept of CCN, which can reduce the content download time and improve the utilization of the network bandwidth.

- **PIT:** The PIT is used to record the passed interest information and sequentially realize that the requested content is smoothly transmitted back to the content requester. The content packet is forwarded back to the content requester step by step according to the records in the PIT. If the content is returned to the requester, the entry is deleted from the PIT.
- **FIB:** The FIB is in charge of sending the interest packet to the destination node. Specially, it can forward the interest packet in multiple directions at the same time.

The CCN forwarding mechanism can be concluded as follows.

- After receiving the request packet, the CS first matches the content cache, if there is relevant content, CS sends it directly, otherwise queries in the PIT.
- If the PIT has a relevant content item, then adds the interest port to the list. In this process, the interest packet is intercepted to prevent duplicate requests for the same content. When there is a content packet response, this content is sent to all ports that request the content.
- If there is no relevant content item in the PIT, then the FIB is queried. The packet is forwarded to the next CCN node as instructed by the FIB.

The advantages of CCN are three-fold:

- The CCN network does not have the security problems of the content channel, because the content can be obtained from any cache.
- The CCN has comparable performance to IP networks in peer-to-peer communication. It has greater flexibility and robustness than IP networks.
- The CCN has the ability to adjust the data traffic. When forwarding a content, the forwarding policy can be selected according to the link status to balance the network traffic.

In vehicular networks, due to the high mobility of vehicles, a group of content replicas of the original content may be kept in different vehicles and infrastructures. As a result, huge overhead is needed to manage these replicas, resulting in low efficiency of content storage and delivery. Therefore, combining the ICN with conventional vehicular networks is an efficient solution to address the above problem. As shown in Fig. 1.8, we have proposed a content centric framework for vehicular networks in [122], where a group of content centric units (CCUs) is distributed in the network to store the replicas of vehicular contents. In the content centric vehicular networks, contents are managed by their naming information, where vehicles can request contents based on their interests. Each CCU has a CS to keep contents, a PIT to manage interests and a FIB to control information routing. With such a network architecture, we have developed a novel algorithm for delivering contents with CCUs, through which the contents can be delivered more efficiently in the content centric vehicular networks.

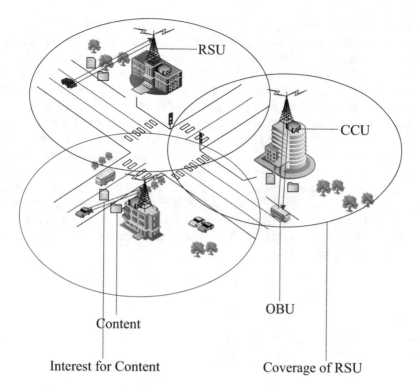

Fig. 1.8 The content centric vehicular networks

1.2.8 Edge Caching

With the development of the Internet of things (IoTs) and communication technologies, various data applications have been an indispensable part of people's life. Along with this, huge mobile traffic is generated from the IoT devices. In current mobile vehicular networks, the massive data demands generated by vehicular users are responded by the remote content provider through infrastructures. There are the following disadvantages.

- Heavy load of backbone networks: the ever-increasing vehicular applications make the mobile backhaul links overload and the bandwidth resources become extremely tense.
- Huge network latency and poor experience for vehicular users: the spatial distance between users and content servers leads to a long latency for the content transmission which decreases the QoE of vehicle users and the revenues of the network operators.

Focus on these problems, edge caching can be a promising solution for both vehicular users and network operators, where the caching services are provided by

MEC servers which are placed at the edge of the networks [123–127]. In general, the remote content servers have the entire network contents and the mobile edge nodes can obtain the contents by connecting to the content servers. Mobile edge nodes are deployed at the edge of the vehicular networks, which are also in close proximity to vehicles. Compared with the remote content servers, the edge nodes have limited storage where only part of the contents can be cached. In order to increase the revenues, the edge nodes usually cache the most popular contents that can be retransmitted to vehicles [128]. If a vehicle intends to request the content that has cached in the edge nodes, the content can be directly transmitted to the vehicle. In this way, the latency for obtaining the required contents can be greatly decreased and the QoE of vehicles can also be improved. In addition, with edge caching, the vehicles do not frequently access the remote content servers with the result that the network load of backbone links can be alleviated and the whole network performance can be optimized [129]. The detailed procedures of the content caching can be seen in Fig. 1.9.

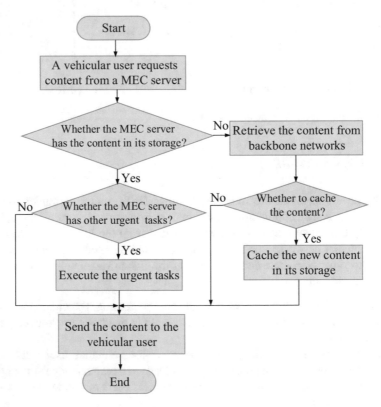

Fig. 1.9 The procedure of content caching

1.2.9 Autonomous Driving

As a novel cyber physical system (CPS), autonomous vehicles (AVs) integrate many functions, such as automotive control, content gathering/caching/processing and communication. Some large scale manufacturers including both car manufacturers (e.g., Audi, BMW, Tesla and Toyota) and Internet companies (e.g., Google and Baidu) declare that the official launching of the first AV to the market is expected in 2020 [130]. It is also predicted that the proportion of AVs on the road will be 25% by 2035.

According to the standard J3016, National Highway Traffic Safety Administration (NHTSA) defines the levels of automation of vehicles:

- **Level 0**—No automation. The majority of current vehicles on the roads belong to this level. In this level, vehicles are full-handled by human drivers. Although some vehicles have certain driver support/convenience systems, they are still considered as vehicles at level 0 because they do not have control authority over steering, braking or throttle, which means that the vehicle driving systems are still controlled by drivers.
- **Level 1**—Function specific automation: The automation at this level involves one or more specific control functions such as steering and braking. However, the control functions are operated independently from each other.
- **Level 2**—Combined function automation: At least two primary control functions that can be operated simultaneously in a vehicle to facilitate the safety driving of vehicles in this level.
- **Level 3**—Limited self-driving automation: In this level, it is not necessary for drivers to monitor the road condition all the time and drivers can let the driving system control the vehicles in some specific conditions. However, this does not mean that the drivers can relax their vigilance because the autonomous control system may malfunction in some complex road conditions.
- **Level 4**—High self-driving automation: In this level, vehicles have higher autonomy, the drivers only need to input the destination or the navigation information, and automated control system will control the vehicle depending on the inputs from sensors and cameras. In this level, drivers do not need to response to the operation requests generated by the autonomous control system.
- **Level 5**—Full self-driving automation: The automated control system is fully in charge of the vehicle and can also gather information (navigation, position, trajectory and intentions) by communicating with other vehicles.

Unfortunately, the AVs with level-5 have not been implemented. This is because the intelligence of a single vehicle cannot fundamentally identify all complex traffic scenarios. For example in 2016, a Tesla MODEL S was using the autopilot function to drive on the road in Florida. Because the collected data was incomplete, the AV could not identify the white truck ahead of it and crashed into the truck directly. In addition, vehicles at high levels, such as level 5, depend heavily on the automated components deployed onboard, e.g, the cameras, the sensors, the navigation systems, etc. Attacks aiming at these components are high threats for the safety driving of AVs. For example, camera and global positioning system (GPS) spoofing/jamming are the two main threats to AVs [131]. Consequently, a vehicle may cause incorrect driving due to its own automated components deployed onboard.

1.2.10   Artificial Intelligence

Artificial intelligence (AI) is a new technical science that studies and develops theories, methods, techniques, and applications for simulating and extending human intelligence [132–134]. It is a branch of computer science that attempts to understand the essence of intelligence and produce a new intelligent machine that responds in a manner similar to human intelligence. Research in this area includes robotics, speech recognition, image recognition, natural language processing and expert systems. Machine learning is a science of AI and can be classified based on the learning style as follows.

- **Supervised learning:** Supervised learning, which is also known as supervised training, refers to the process of using a set of known categories of samples to train the parameters of the classifier with the target of achieving the required performance. Specifically, supervised learning is a machine learning task that infers a function from the labeled training data, where the training data includes a set of training examples. In supervised learning, each instance consists of an input object (usually a vector) and a desired output value (also known as a supervisory signal). The supervised learning algorithm analyzes the training data and produces an inferred function that can be used to map out new instances.
- **Unsupervised learning:** Unlike supervised learning, unsupervised learning has no labeled training data. This is due to the lack of sufficient prior knowledge or the consideration of label cost. Thus, it is difficult to manually label categories. Consequently, the unsupervised learning aims to solve various problems by automatically extracting the features through the well designed algorithms.
- **Reinforcement learning:** Reinforcement learning, also known as enhanced learning, is the way in which the agent learns in a trial and error manner. The behavior of the agent can be guided by interacting with the environment with the target of obtaining the biggest reward. Unlike supervised learning, the enhanced signal provided by the environment in reinforcement learning is an evaluation of the quality of the action (usually a scalar signal) rather than telling the reinforcement learning system how to produce the correct action. In this way, reinforcement learning gains knowledge in an action-evaluation environment and improves the action plan to adapt to the environment.

Deep learning is a new research direction in the field of machine learning [135–139]. It enables the machine to imitate human activities such as audiovisual and thinking. The information obtained in the learning processes is expected to enable machines to analyze and learn like humans, and to recognize data such as texts, images and sounds. Deep learning is a complex machine learning algorithm that has achieved a lot in search technology, data mining, machine learning, machine translation, natural language processing, multimedia learning, recommendation and personalization technology, and other related fields.

Based on the advantages of deep learning, we can know that it can significantly improve the vehicle's ability to detect and classify the surrounding objects, thus facilitating the fusion of data obtained by different sensors [140–145]. By analyzing the massive data generated by city transportation through data analysis and deep learning technologies, the in-depth understanding of road traffic demand and network state can be provided for decision-makers. To be specific, AVs can recognize surrounding environment, classify information and predict the motion of objects by inputting the perceived data into the trained learning model to make driving decisions. The large-scale mobility data generated by vehicles can be analyzed and used by the cloud server to guide traffic scheduling and network infrastructure deployment. Being faced with the complex applications in the vehicular networks, different learning algorithms trained by different data will provide various performances. The wrong results obtained from learning models will lead to a great loss of life and property. In addition, different learning algorithms are suitable for different applications and require different training time. Designing the optimal learning algorithm for a specific traffic application therefore becomes a challenge that needs the attention from both academia and industry.

1.3   Aim of the Book

The next generation vehicular networks target to provide vehicular services with highly improved network performance in terms of decreasing the service cost and latency while improving the network efficiency and QoE of vehicles. Towards better network maintainability and sustainability, this book aims to propose the novel network envisions and framework design principles to realize the goals of the next generation vehicular networks. As is widely recognized, enabling technologies such as ICN, edge caching, computation offloading, deep learning and autonomous driving play important roles in the wireless technology developments along with the research on the ITS and smart cities. The investigation and development on the integration of vehicular networks and the new technologies therefore becomes a new research direction and can provide important performance references for network framework design and future standardizations.

This book will focus on the key enabling technologies to design the framework for the next generation vehicular networks to satisfy various vehicular applications and services requested by vehicles. Based on the most recent research results of the enabling technologies from academia and industry, this book will describe the integration of conventional vehicular networks and the enabling technologies in depth to satisfy the goals of the next generation vehicular networks. The network architectures and the framework design principles will be discussed in depth including the analysis of reputation based content delivery in information centric vehicular networks, contract based edge caching in vehicular networks, Stackelberg game based computation offloading in vehicular networks, auction game based secure computation offloading in vehicular networks, bargain game based secure content delivery in vehicular networks and deep learning based autonomous driving in vehicular networks, respectively. Based on the investigations on the integration of vehicular networks and the enabling technologies, it can help us elaborate the insights and implications to design and implement the next generation vehicular networks.

References

1. Z. Xiao, X. Shen, F. Zeng, V. Havyarimana, D. Wang, W. Chen, K. Li, Spectrum resource sharing in heterogeneous vehicular networks: a noncooperative game-theoretic approach with correlated equilibrium. IEEE Trans. Veh. Technol. **67**(10), 9449–9458 (2018)
2. T. Wang, X. Cao, S. Wang, Self-adaptive clustering and load-bandwidth management for uplink enhancement in heterogeneous vehicular networks. IEEE Internet Things J. **6**(3), 5607–5617 (2019)
3. Y. Hui, Z. Su, T.H. Luan, J. Cai, A game theoretic scheme for optimal access control in heterogeneous vehicular networks. IEEE Trans. Intell. Transp. Syst. **20**(12), 4590–4603 (2019)
4. P. Dai, K. Liu, X. Wu, Y. Liao, V.C.S. Lee, S.H. Son, Bandwidth efficiency and service adaptiveness oriented data dissemination in heterogeneous vehicular networks. IEEE Trans. Veh. Technol. **67**(7), 6585–6598 (2018)
5. X. Zhao, X. Li, Z. Xu, T. Chen, An optimal game approach for heterogeneous vehicular network selection with varying network performance. IEEE Intell. Transp. Syst. Mag. **11**(3), 80–92 (2019)
6. W. Xu, W. Shi, F. Lyu, H. Zhou, N. Cheng, X. Shen, Throughput analysis of vehicular internet access via roadside wifi hotspot. IEEE Trans. Veh. Technol. **68**(4), 3980–3991 (2019)
7. L. Liang, H. Ye, G.Y. Li, Toward intelligent vehicular networks: a machine learning framework. IEEE Internet Things J. **6**(1), 124–135 (2019)
8. Y. Hui, Z. Su, S. Guo, Utility based data computing scheme to provide sensing service in internet of things. IEEE Trans. Emerg. Top. Comput. **7**(2), 337–348 (2019)
9. Z. Zhou, J. Feng, Z. Chang, X. Shen, Energy-efficient edge computing service provisioning for vehicular networks: a consensus admm approach. IEEE Trans. Veh. Technol. **68**(5), 5087–5099 (2019)
10. H. Peng, L. Liang, X. Shen, G.Y. Li, Vehicular communications: a network layer perspective. IEEE Trans. Veh. Technol. **68**(2), 1064–1078 (2019)
11. M.A. Togou, L. Khoukhi, A. Hafid, Performance analysis and enhancement of wave for v2v non-safety applications. IEEE Trans. Intell. Transp. Syst. **19**(8), 2603–2614 (2018)
12. S. Darbha, S. Konduri, P.R. Pagilla, Benefits of v2v communication for autonomous and connected vehicles. IEEE Trans. Intell. Transp. Syst. **20**(5), 1954–1963 (2019)
13. J. Mei, K. Zheng, L. Zhao, Y. Teng, X. Wang, A latency and reliability guaranteed resource allocation scheme for lte v2v communication systems. IEEE Trans. Wireless Commun. **17**(6), 3850–3860 (2018)
14. F. Abbas, P. Fan, Z. Khan, A novel low-latency v2v resource allocation scheme based on cellular v2x communications. IEEE Trans. Intell. Transp. Syst. **20**(6), 2185–2197 (2019)

15. P.S. Bithas, A.G. Kanatas, D.B. da Costa, P.K. Upadhyay, U.S. Dias, On the double-generalized gamma statistics and their application to the performance analysis of v2v communications. IEEE Trans. Commun. **66**(1), 448–460 (2018)
16. R. Atallah, M. Khabbaz, C. Assi, Multihop v2i communications: a feasibility study, modeling, and performance analysis. IEEE Trans. Veh. Technol. **66**(3), 2801–2810 (2017)
17. O. Popescu, S. Sha-Mohammad, H. Abdel-Wahab, D.C. Popescu, S. El-Tawab, Automatic incident detection in intelligent transportation systems using aggregation of traffic parameters collected through v2i communications. IEEE Intell. Transp. Syst. Mag. **9**(2), 64–75 (2017)
18. Z. Su, Y. Hui, T.H. Luan, S. Guo, Engineering a game theoretic access for urban vehicular networks. IEEE Trans. Veh. Technol. **66**(6), 4602–4615 (2017)
19. J. Shi, Z. Yang, H. Xu, M. Chen, B. Champagne, Dynamic resource allocation for lte-based vehicle-to-infrastructure networks. IEEE Trans. Veh. Technol. **68**(5), 5017–5030 (2019)
20. F. Jiang, C. Li, Z. Gong, Low complexity and fast processing algorithms for v2i massive mimo uplink detection. IEEE Trans. Veh. Technol. **67**(6), 5054–5068 (2018)
21. A. Boualouache, S. Senouci, S. Moussaoui, A survey on pseudonym changing strategies for vehicular ad-hoc networks. IEEE Commun. Surv. Tutorials **20**(1), 770–790 (2018)
22. P.S. Bithas, G.P. Efthymoglou, A.G. Kanatas, V2V cooperative relaying communications under interference and outdated CSI. IEEE Trans. Veh. Technol. **67**(4), 3466–3480 (2018)
23. Z. Su, Y. Hui, S. Guo, D2d-based content delivery with parked vehicles in vehicular social networks. IEEE Wirel. Commun. **23**(4), 90–95 (2016)
24. D.M. Mughal, J.S. Kim, H. Lee, M.Y. Chung, Performance analysis of v2v communications: a novel scheduling assignment and data transmission scheme. IEEE Trans. Veh. Technol. **68**(7), 7045–7056 (2019)
25. J. Gao, M. Li, L. Zhao, X. Shen, Contention intensity based distributed coordination for v2v safety message broadcast. IEEE Trans. Veh. Technol. **67**(12), 12288–12301 (2018)
26. H. Yao, D. Zeng, H. Huang, S. Guo, A. Barnawi, I. Stojmenovic, Opportunistic offloading of deadline-constrained bulk cellular traffic in vehicular DTNs. IEEE Trans. Comput. **64**(12), 3515–3527 (2015)
27. P. Kolios, V. Friderikos, K. Papadaki, Energy-efficient relaying via store-carry and forward within the cell. IEEE Trans. Mobile Comput. **13**(1), 202–215 (2014)
28. J. He, L. Cai, J. Pan, P. Cheng, Delay analysis and routing for two-dimensional vanets using carry-and-forward mechanism. IEEE Trans. Mobile Comput. **16**(7), 1830–1841 (2017)
29. Q. Xu, Z. Su, K. Zhang, P. Ren, X. Shen, Epidemic information dissemination in mobile social networks with opportunistic links. IEEE Trans. Emerg. Top. Comput. **3**(3), 399–409 (2015)
30. K. Zheng, L. Hou, H. Meng, Q. Zheng, N. Lu, L. Lei, Soft-defined heterogeneous vehicular network: architecture and challenges. IEEE Netw. **30**(4), 72–80 (2016)
31. Z. He, J. Cao, X. Liu, SDVN: enabling rapid network innovation for heterogeneous vehicular communication. IEEE Netw. **30**(4), 10–15 (2016)
32. Y. Hui, Z. Su, T.H. Luan, Collaborative content delivery in software-defined heterogeneous vehicular networks. IEEE/ACM Trans. Netw. **28**(2), 575–587 (2020)
33. K. Zheng, Q. Zheng, P. Chatzimisios, W. Xiang, Y. Zhou, Heterogeneous vehicular networking: a survey on architecture, challenges, and solutions. IEEE Commun. Surv. Tutorials **17**(4), 2377–2396 (2015)
34. M. Xing, J. He, L. Cai, Utility maximization for multimedia data dissemination in large-scale vanets. IEEE Trans. Mobile Comput. **16**(4), 1188–1198 (2017)
35. J. Qiao, Y. He, X.S. Shen, Improving video streaming quality in 5g enabled vehicular networks. IEEE Wirel. Commun. **25**(2), 133–139 (2018)
36. J. Guo, B. Song, Y. He, F.R. Yu, M. Sookhak, A survey on compressed sensing in vehicular infotainment systems. IEEE Commun. Surv. Tutorials **19**(4), 2662–2680 (2017)
37. L. Sarakis, T. Orphanoudakis, H.C. Leligou, S. Voliotis, A. Voulkidis, Providing entertainment applications in vanet environments. IEEE Wirel. Commun. **23**(1), 30–37 (2016)

38. E. Costa-Montenegro, F. Quinoy-Garcia, F.J. Gonzalez-castano, F. Gil-Castineira, Vehicular entertainment systems: mobile application enhancement in networked infrastructures. IEEE Veh. Technol. Mag. **7**(3), 73–79 (2012)
39. C. Wang, Y. Li, D. Jin, S. Chen, On the serviceability of mobile vehicular cloudlets in a large-scale urban environment. IEEE Trans. Intell. Transp. Syst. **17**(10), 2960–2970 (2016)
40. T. ETSI, Intelligent transport systems (its); vehicular communications; basic set of applications; definitions, Tech. Rep. ETSI TR 102 638, Tech. Rep., 2009
41. E. Smith, Statistics on intersection accidents, https://www.autoaccident.com/statistics-on-intersection-accidents.html
42. F.J. Martinez, C.K. Toh, J.C. Cano, C.T. Calafate, P. Manzoni, Emergency services in future intelligent transportation systems based on vehicular communication networks. IEEE Intell. Transp. Syst. Mag. **2**(2), 6–20 (2010)
43. L. Wang, T. Han, Q. Li, J. Yan, X. Liu, D. Deng, Cell-less communications in 5g vehicular networks based on vehicle-installed access points. IEEE Wirel. Commun. **24**(6), 64–71 (2017)
44. J. Nightingale, P. Salva-Garcia, J.M.A. Calero, Q. Wang, 5g-QoE: QoE modelling for ultra-hd video streaming in 5g networks. IEEE Trans. Broadcast. **64**(2), 621–634 (2018)
45. C. Mao, M. Khalily, P. Xiao, T.W.C. Brown, S. Gao, Planar sub-millimeter-wave array antenna with enhanced gain and reduced sidelobes for 5g broadcast applications. IEEE Trans. Antennas Propag. **67**(1), 160–168 (2019)
46. V. Petrov, M.A. Lema, M. Gapeyenko, K. Antonakoglou, D. Moltchanov, F. Sardis, A. Samuylov, S. Andreev, Y. Koucheryavy, M. Dohler, Achieving end-to-end reliability of mission-critical traffic in softwarized 5g networks. IEEE J. Sel. Areas Commun. **36**(3), 485–501 (2018)
47. T.K. Vu, M. Bennis, M. Debbah, M. Latva-Aho, Joint path selection and rate allocation framework for 5g self-backhauled mm-wave networks. IEEE Trans. Wireless Commun. **18**(4), 2431–2445 (2019)
48. W. Lu, X. Meng, G. Guo, Fast service migration method based on virtual machine technology for MEC. IEEE Internet Things J. **6**(3), 4344–4354 (2019)
49. X. He, R. Jin, H. Dai, Deep PDS-learning for privacy-aware offloading in MEC-enabled IoT. IEEE Internet Things J. **6**(3), 4547–4555 (2019)
50. Z. Ding, P. Fan, H.V. Poor, Impact of non-orthogonal multiple access on the offloading of mobile edge computing. IEEE Trans. Commun. **67**(1), 375–390 (2019)
51. Z. Ning, P. Dong, X. Kong, F. Xia, A cooperative partial computation offloading scheme for mobile edge computing enabled internet of things. IEEE Internet Things J. **6**(3), 4804–4814 (2019)
52. J. Zhang, X. Hu, Z. Ning, E.C. Ngai, L. Zhou, J. Wei, J. Cheng, B. Hu, V.C.M. Leung, Joint resource allocation for latency-sensitive services over mobile edge computing networks with caching. IEEE Internet Things J. **6**(3), 4283–4294 (2019)
53. T.Q. Dinh, Q.D. La, T.Q.S. Quek, H. Shin, Learning for computation offloading in mobile edge computing. IEEE Trans. Commun. **66**(12), 6353–6367 (2018)
54. X. Lyu, W. Ni, H. Tian, R.P. Liu, X. Wang, G.B. Giannakis, A. Paulraj, Optimal schedule of mobile edge computing for internet of things using partial information. IEEE J. Sel. Areas Commun. **35**(11), pp. 2606–2615 (2017)
55. S. Sardellitti, G. Scutari, S. Barbarossa, Joint optimization of radio and computational resources for multicell mobile-edge computing. IEEE Trans. Signal Inf. Process. Netw. **1**(2), 89–103 (2015)
56. X. Chen, L. Jiao, W. Li, X. Fu, Efficient multi-user computation offloading for mobile-edge cloud computing. IEEE/ACM Trans. Netw. **24**(5), 2795–2808 (2016)
57. A. Fischer, J.F. Botero, M.T. Beck, H. de Meer, X. Hesselbach, Virtual network embedding: a survey. IEEE Commun. Surv. Tutorials **15**(4), 1888–1906 (2013)
58. V.G. Nguyen, A. Brunstrom, K.J. Grinnemo, J. Taheri, SDN/NFV-based mobile packet core network architectures: a survey. IEEE Commun. Surv. Tutorials **19**(3), 1567–1602 (2017)
59. X. Cheng, Y. Wu, G. Min, A.Y. Zomaya, Network function virtualization in dynamic networks: a stochastic perspective. IEEE J. Sel. Areas Commun. **36**(10), 2218–2232 (2018)

60. R. Mijumbi, J. Serrat, J. Gorricho, N. Bouten, F. De Turck, R. Boutaba, Network function virtualization: state-of-the-art and research challenges. IEEE Commun. Surv. Tutorials **18**(1), 236–262 (2016)

61. D. Cotroneo, R. Natella, S. Rosiello, NFV-throttle: an overload control framework for network function virtualization. IEEE Trans. Netw. Serv. Manag. **14**(4), 949–963 (2017).

62. R. Mijumbi, J. Serrat, J.L. Gorricho, N. Bouten, F.D. Turck, R. Boutaba, Network function virtualization: state-of-the-art and research challenges. IEEE Commun. Surv. Tutorials **18**(1), 236–262 (2015)

63. B. Han, V. Gopalakrishnan, L. Ji, S. Lee, Network function virtualization: Challenges and opportunities for innovations. IEEE Commun. Mag. **53**(2), 90–97 (2015)

64. T. Taleb, K. Samdanis, B. Mada, H. Flinck, S. Dutta, D. Sabella, On multi-access edge computing: a survey of the emerging 5g network edge cloud architecture and orchestration. IEEE Commun. Surv. Tutorials **19**(3), 1657–1681 (2017)

65. R. Riggio, A. Bradai, D. Harutyunyan, T. Rasheed, T. Ahmed, Scheduling wireless virtual networks functions. IEEE Trans. Netw. Serv. Manage. **13**(2), 240–252 (2016)

66. M. Zhu, J. Cao, Z. Cai, Z. He, M. Xu, Providing flexible services for heterogeneous vehicles: an NFV-based approach. IEEE Netw. **30**(3), 64–71 (2016)

67. S. Khan, A. Gani, A.W.A. Wahab, M. Guizani, M.K. Khan, Topology discovery in software defined networks: threats, taxonomy, and state-of-the-art. IEEE Commun. Surv. Tutorials **19**(1), 303–324 (2016)

68. S. Khan, A. Gani, A.W.A. Wahab, A. Abdelaziz, K. Ko, M.K. Khan, M. Guizani, Software-defined network forensics: motivation, potential locations, requirements, and challenges. IEEE Netw. **30**(6), 6–13 (2016)

69. M.A. Salahuddin, A. Al-Fuqaha, M. Guizani, Software-defined networking for rsu clouds in support of the internet of vehicles. IEEE Internet Things J. **2**(2), 133–144 (2015)

70. R. Jain, S. Paul, Network virtualization and software defined networking for cloud computing: a survey. IEEE Commun. Mag. **51**(11), 24–31 (2013)

71. D. Kreutz, F.M.V. Ramos, P.E. Verĺssimo, C.E. Rothenberg, S. Azodolmolky, S. Uhlig, Software-defined networking: a comprehensive survey. Proc. IEEE **103**(1), 14–76 (2015)

72. S. Garg, K. Kaur, S.H. Ahmed, A. Bradai, G. Kaddoum, M. Atiquzzaman, MobQoS: Mobility-aware and QoS-driven SDN framework for autonomous vehicles. IEEE Wirel. Commun. **26**(4), 12–20 (2019)

73. R. Amin, M. Reisslein, N. Shah, Hybrid SDN networks: a survey of existing approaches. IEEE Commun. Surv. Tutorials **20**(4), 3259–3306 (2018)

74. G. Yu, R. Liu, Q. Chen, Z. Tang, A hierarchical sdn architecture for ultra-dense millimeter-wave cellular networks. IEEE Commun. Mag. **56**(6), 79–85 (2018)

75. Z. Su, Q. Xu, H. Zhu, Y. Wang, A novel design for content delivery over software defined mobile social networks. IEEE Netw. **29**(4), 62–67 (2015)

76. K. Wang, Y. Wang, D. Zeng, S. Guo, An SDN-based architecture for next-generation wireless networks. IEEE Wirel. Commun. **24**(1), 25–31 (2017)

77. H. Li, M. Dong, K. Ota, Control plane optimization in software-defined vehicular ad hoc networks. IEEE Trans. Veh. Technol. **65**(10), 7895–7904 (2016)

78. J. Weng, J. Weng, Y. Zhang, W. Luo, W. Lan, BENBI: scalable and dynamic access control on the northbound interface of SDN-based vanet. IEEE Trans. Veh. Technol. **68**(1), 822–831 (2019)

79. K. Liu, L. Feng, P. Dai, V.C.S. Lee, S.H. Son, J. Cao, Coding-assisted broadcast scheduling via memetic computing in SDN-based vehicular networks. IEEE Trans. Intell. Transp. Syst. **19**(8), 2420–2431 (2018)

80. J. Liu, J. Wan, B. Zeng, Q. Wang, H. Song, M. Qiu, A scalable and quick-response software defined vehicular network assisted by mobile edge computing. IEEE Commun. Mag. **55**(7), 94–100 (2017)

81. X. Huang, R. Yu, J. Kang, Z. Xia, Y. Zhang, Software defined networking for energy harvesting internet of things. IEEE Internet Things J. **5**(3), 1389–1399 (2018)

82. A. Lara, A. Kolasani, B. Ramamurthy, Network innovation using openflow: a survey. IEEE Commun. Surv. Tutorials **16**(1), 493–512 (2013)

83. C.J. Bernardos, A. de la Oliva, P. Serrano, A. Banchs, L.M. Contreras, H. Jin, J.C. Zuniga, An architecture for software defined wireless networking. IEEE Wirel. Commun. **21**(3), 52–61 (2014)

84. F. Hu, Q. Hao, K. Bao, A survey on software-defined network and openflow: From concept to implementation. IEEE Commun. Surv. Tutorials **16**(4), 2181–2206 (2014)

85. J. Chen, H. Zhou, N. Zhang, W. Xu, Q. Yu, L. Gui, X. Shen, Service-oriented dynamic connection management for software-defined internet of vehicles. IEEE Trans. Intell. Transp. Syst. **18**(10), 2826–2837 (2017)

86. C. Wang, C. Liang, F.R. Yu, Q. Chen, L. Tang, Computation offloading and resource allocation in wireless cellular networks with mobile edge computing. IEEE Trans. Wireless Commun. **16**(8), 4924–4938 (2017)

87. J. Zhao, Q. Li, Y. Gong, K. Zhang, Computation offloading and resource allocation for cloud assisted mobile edge computing in vehicular networks. IEEE Trans. Veh. Technol. **68**(8), 7944–7956 (2019)

88. J. Du, F.R. Yu, X. Chu, J. Feng, G. Lu, Computation offloading and resource allocation in vehicular networks based on dual-side cost minimization. IEEE Trans. Veh. Technol. **68**(2), 1079–1092 (2019)

89. Y. Wu, L.P. Qian, H. Mao, X. Yang, H. Zhou, X. Tan, D.H.K. Tsang, Secrecy-driven resource management for vehicular computation offloading networks. IEEE Netw. **32**(3), 84–91 (2018)

90. Z. Su, Y. Hui, T.H. Luan, Distributed task allocation to enable collaborative autonomous driving with network softwarization. IEEE J. Sel. Areas Commun. **36**(10), 2175–2189 (2018)

91. X. Hou, Y. Li, M. Chen, D. Wu, D. Jin, S. Chen, Vehicular fog computing: a viewpoint of vehicles as the infrastructures. IEEE Trans. Veh. Technol. **65**(6), 3860–3873 (2016)

92. B. Brik, N. Lagraa, N. Tamani, A. Lakas, Y. Ghamri-Doudane, Renting out cloud services in mobile vehicular cloud. IEEE Trans. Veh. Technol. **67**(10), 9882–9895 (2018)

93. E. Lee, E. Lee, M. Gerla, S.Y. Oh, Vehicular cloud networking: architecture and design principles. IEEE Commun. Mag. **52**(2), 148–155 (2014)

94. S. Wang, J. Wang, X. Wang, T. Qiu, Y. Yuan, L. Ouyang, Y. Guo, F. Wang, Blockchain-powered parallel healthcare systems based on the acp approach. IEEE Trans. Comput. Soc. Syst. **5**(4), 942–950 (2018)

95. D. Liu, A. Alahmadi, J. Ni, X. Lin, X. Shen, Anonymous reputation system for IIoT-enabled retail marketing atop PoS blockchain. IEEE Trans. Ind. Inf. **15**(6), 3527–3537 (2019)

96. P. Danzi, A.E. Kalør, Č. Stefanović, P. Popovski, Delay and communication tradeoffs for blockchain systems with lightweight IoT clients. IEEE Internet Things J. **6**(2), 2354–2365 (2019)

97. M. Liu, F.R. Yu, Y. Teng, V.C.M. Leung, M. Song, Performance optimization for blockchain-enabled industrial internet of things (IIoT) systems: a deep reinforcement learning approach. IEEE Trans. Ind. Inf. **15**(6), 3559–3570 (2019)

98. Y. Sun, L. Zhang, G. Feng, B. Yang, B. Cao, M.A. Imran, Blockchain-enabled wireless internet of things: performance analysis and optimal communication node deployment. IEEE Internet Things J. **6**(3), 5791–5802 (2019)

99. H. Yao, T. Mai, J. Wang, Z. Ji, C. Jiang, Y. Qian, Resource trading in blockchain-based industrial internet of things. IEEE Trans. Ind. Inf. **15**(6), 3602–3609 (2019)

100. J. Wan, J. Li, M. Imran, D. Li, A blockchain-based solution for enhancing security and privacy in smart factory. IEEE Trans. Ind. Inf. **15**(6), 3652–3660 (2019)

101. J. Huang, L. Kong, G. Chen, M. Wu, X. Liu, P. Zeng, Towards secure industrial IoT: Blockchain system with credit-based consensus mechanism. IEEE Trans. Ind. Inf. **15**(6), 3680–3689 (2019)

102. Y. Zhang, S. Kasahara, Y. Shen, X. Jiang, J. Wan, Smart contract-based access control for the internet of things. IEEE Internet Things J. **6**(2), 1594–1605 (2019)

103. Z. Su, Y. Wang, Q. Xu, M. Fei, Y. Tian, N. Zhang, A secure charging scheme for electric vehicles with smart communities in energy blockchain. IEEE Internet Things J. **6**(3), 4601–4613 (2019)

104. J. Pan, J. Wang, A. Hester, I. Alqerm, Y. Liu, Y. Zhao, Edgechain: an edge-IoT framework and prototype based on blockchain and smart contracts. IEEE Internet Things J. **6**(3), 4719–4732 (2019)

105. Z. Yang, K. Yang, L. Lei, K. Zheng, V.C.M. Leung, Blockchain-based decentralized trust management in vehicular networks. IEEE Internet Things J. **6**(2), 1495–1505 (2019)

106. M. Li, L. Zhu, X. Lin, Efficient and privacy-preserving carpooling using blockchain-assisted vehicular fog computing. IEEE Internet Things J. **6**(3), 4573–4584 (2019)

107. T. Jiang, H. Fang, H. Wang, Blockchain-based internet of vehicles: distributed network architecture and performance analysis. IEEE Internet Things J. **6**(3), 4640–4649 (2019)

108. Y. Wang, Z. Su, N. Zhang, BSIS: blockchain-based secure incentive scheme for energy delivery in vehicular energy network. IEEE Trans. Ind. Inf. **15**(6), 3620–3631 (2019)

109. J. Kang, R. Yu, X. Huang, M. Wu, S. Maharjan, S. Xie, Y. Zhang, Blockchain for secure and efficient data sharing in vehicular edge computing and networks. IEEE Internet Things J. **6**(3), 4660–4670 (2019)

110. V. Ortega, F. Bouchmal, J.F. Monserrat, Trusted 5g vehicular networks: blockchains and content-centric networking. IEEE Veh. Technol. Mag. **13**(2), 121–127 (2018)

111. C. Xu, M. Wang, X. Chen, L. Zhong, L.A. Grieco, Optimal information centric caching in 5g device-to-device communications. IEEE Trans. Mobile Comput. **17**(9), 2114–2126 (2018)

112. Y. Zhou, F.R. Yu, J. Chen, Y. Kuo, Resource allocation for information-centric virtualized heterogeneous networks with in-network caching and mobile edge computing. IEEE Trans. Veh. Technol. **66**(12), 11339–11351 (2017)

113. K. Xu, Y. Wan, G. Xue, Powering smart homes with information-centric networking. IEEE Commun. Mag. **57**(6), 40–46 (2019)

114. H. Yao, M. Li, J. Du, P. Zhang, C. Jiang, Z. Han, Artificial intelligence for information-centric networks. IEEE Commun. Mag. **57**(6), 47–53 (2019)

115. C. Liang, F.R. Yu, H. Yao, Z. Han, Virtual resource allocation in information-centric wireless networks with virtualization. IEEE Trans. Veh. Technol. **65**(12), 9902–9914 (2016)

116. G. Xylomenos, C.N. Ververidis, V.A. Siris, N. Fotiou, C. Tsilopoulos, X. Vasilakos, K.V. Katsaros, G.C. Polyzos, A survey of information-centric networking research. IEEE Commun. Surv. Tutorials **16**(2), 1024–1049 (2014)

117. R. Wang, X. Peng, J. Zhang, K.B. Letaief, Mobility-aware caching for content-centric wireless networks: modeling and methodology. IEEE Commun. Mag. **54**(8), 77–83 (2016)

118. H. Asaeda, K. Matsuzono, T. Turletti, Contrace: a tool for measuring and tracing content-centric networks. IEEE Commun. Mag. **53**(3), 182–188 (2015)

119. Z. Su, Q. Xu, Content distribution over content centric mobile social networks in 5g. IEEE Commun. Mag. **53**(6), 66–72 (2015)

120. Q. Wu, Z. Li, G. Tyson, S. Uhlig, M.A. Kaafar, G. Xie, Privacy-aware multipath video caching for content-centric networks. IEEE J. Sel. Areas Commun. **34**(8), 2219–2230 (2016)

121. T. Semertzidis, P. Daras, P. Moore, L. Makris, M.G. Strintzis, Automatic creation of 3d environments from a single sketch using content-centric networks. IEEE Commun. Mag. **49**(3), 152–157 (2011)

122. Z. Su, Y. Hui, Q. Yang, The next generation vehicular networks: a content-centric framework. IEEE Wirel. Commun. **24**(1), 60–66 (2017)

123. A. Mahmood, C.E. Casetti, C.F. Chiasserini, P. Giaccone, J. Harri, The rich prefetching in edge caches for in-order delivery to connected cars. IEEE Trans. Veh. Technol. **68**(1), 4–18 (2019)

124. Z. Su, Y. Hui, Q. Xu, T. Yang, J. Liu, Y. Jia, An edge caching scheme to distribute content in vehicular networks. IEEE Trans. Veh. Technol. **67**(6), 5346–5356 (2018)

125. L.T. Tan, R.Q. Hu, L. Hanzo, Twin-timescale artificial intelligence aided mobility-aware edge caching and computing in vehicular networks. IEEE Trans. Veh. Technol. **68**(4), 3086–3099 (2019)

126. Y. Hui, Z. Su, T.H. Luan, J. Cai, Content in motion: an edge computing based relay scheme for content dissemination in urban vehicular networks. IEEE Trans. Intell. Transp. Syst. **20**(8), 3115–3128 (2019)

127. K. Zhang, S. Leng, Y. He, S. Maharjan, Y. Zhang, Cooperative content caching in 5g networks with mobile edge computing. IEEE Wirel. Commun. **25**(3), 80–87 (2018)

128. Q. Xu, Z. Su, Q. Zheng, M. Luo, B. Dong, Secure content delivery with edge nodes to save caching resources for mobile users in green cities. IEEE Trans. Ind. Inf. **14**(6), 2550–2559 (2018)

129. E. Bastug, M. Bennis, M. Debbah, Living on the edge: the role of proactive caching in 5g wireless networks. IEEE Commun. Mag. **52**(8), 82–89 (2014)

130. N. Li, D.W. Oyler, M. Zhang, Y. Yildiz, I. Kolmanovsky, A.R. Girard, Game theoretic modeling of driver and vehicle interactions for verification and validation of autonomous vehicle control systems. IEEE Trans. Control Syst. Technol. **26**(5), 1782–1797 (2018)

131. J. Petit, S.E. Shladover, Potential cyberattacks on automated vehicles. IEEE Trans. Intell. Transp. Syst. **16**(2), 546–556 (2015)

132. Z. Zhou, X. Chen, E. Li, L. Zeng, K. Luo, J. Zhang, Edge intelligence: paving the last mile of artificial intelligence with edge computing. Proc. IEEE **107**(8),1738–1762 (2019)

133. L. Li, N. Zheng, F. Wang, On the crossroad of artificial intelligence: a revisit to alan turing and norbert wiener. IEEE Trans. Cybern. **49**(10), 3618–3626 (2019)

134. G. Acampora, D.J. Cook, P. Rashidi, A.V. Vasilakos, A survey on ambient intelligence in healthcare. Proc. IEEE **101**(12), 2470–2494 (2013)

135. S. Hussein, P. Kandel, C.W. Bolan, M.B. Wallace, U. Bagci, Lung and pancreatic tumor characterization in the deep learning era: novel supervised and unsupervised learning approaches. IEEE Trans. Med. Imaging **38**(8), 1777–1787 (2019)

136. L. Shao, D. Wu, X. Li, Learning deep and wide: a spectral method for learning deep networks. IEEE Trans. Neural Netw. Learn. Syst. **25**(12), 2303–2308 (2014)

137. M. Mahmud, M.S. Kaiser, A. Hussain, S. Vassanelli, Applications of deep learning and reinforcement learning to biological data. IEEE Trans. Neural Netw. Learn. Syst. **29**(6), 2063–2079 (2018)

138. Z. Chen, L. Duan, S. Wang, Y. Lou, T. Huang, D.O. Wu, W. Gao, Toward knowledge as a service over networks: a deep learning model communication paradigm. IEEE J. Sel. Areas Commun. **37**(6), 1349–1363 (2019)

139. Z.M. Fadlullah, F. Tang, B. Mao, N. Kato, O. Akashi, T. Inoue, K. Mizutani, State-of-the-art deep learning: evolving machine intelligence toward tomorrows intelligent network traffic control systems. IEEE Commun. Surv. Tutorials **19**(4), 2432–2455 (2017)

140. Q. Wang, J. Wan, X. Li, Robust hierarchical deep learning for vehicular management. IEEE Trans. Veh. Technol. **68**(5), 4148–4156 (2019)

141. Q. Qi, J. Wang, Z. Ma, H. Sun, Y. Cao, L. Zhang, J. Liao, Knowledge-driven service offloading decision for vehicular edge computing: a deep reinforcement learning approach. IEEE Trans. Veh. Technol. **68**(5), 4192–4203 (2019)

142. R.F. Atallah, C.M. Assi, M.J. Khabbaz, Scheduling the operation of a connected vehicular network using deep reinforcement learning. IEEE Trans. Intell. Transp. Syst. **20**(5), 1669–1682 (2019)

143. X. Liang, X. Du, G. Wang, Z. Han, A deep reinforcement learning network for traffic light cycle control. IEEE Trans. Veh. Technol. **68**(2), 1243–1253 (2019)

144. Y. He, N. Zhao, H. Yin, Integrated networking, caching, and computing for connected vehicles: a deep reinforcement learning approach. IEEE Trans. Veh. Technol. **67**(1), 44–55 (2018)

145. Y. Wang, M. Liu, J. Yang, G. Gui, Data-driven deep learning for automatic modulation recognition in cognitive radios. IEEE Trans. Veh. Technol. **68**(4), 4074–4077 (2019)

Chapter 2
Reputation Based Content Delivery in Information Centric Vehicular Networks

The information centric vehicular networks, which integrate the information centric networking (ICN) and the vehicular networks, have advocated recently as a key enabling technology to provide vehicles with efficient content delivery services. However, due to the ever-increasing number of vehicles and their various demands for quality of experience (QoE), the problem of trust for content delivery in information centric vehicular networks becomes a new challenge. To this end, in this chapter, a reputation based framework to deliver content in information centric vehicular networks is presented. Specifically, we first design a graphic user interface (GUI) for each vehicle to manage its interests and reputation value. Then, the methods to calculate the reputation of vehicles and the costs to pay for the content delivery services are shown, respectively. With the incentives, vehicles are motivated to take part in the content delivery with the target of increasing their reputation and earning profits. After this, the update of vehicles' reputation is designed and a Bayes based scheme is used to divide the vehicles into two groups which are trustworthy and untrustworthy, respectively. Finally, a simulation experiment is carried out to demonstrate the efficiency of the proposed framework with comparisons to the existing scheme.

2.1 Introduction

The explosive growth of various content delivery services requested by vehicles has induced the investigation of the vehicular networks [1–5]. In the vehicular networks, as shown in Fig. 2.1, there are base stations (BSs) and roadside units (RSUs), where the RSUs typically have lower transmission power compared to the BSs [6]. Both the BSs and RSUs are placed along roads to provide safety and non-safety contents to vehicles by using the vehicle to infrastructures (V2I) communication [7–11].

© Springer Nature Switzerland AG 2021
Z. Su et al., *The Next Generation Vehicular Networks, Modeling, Algorithm, and Applications*, Wireless Networks, https://doi.org/10.1007/978-3-030-56827-6_2

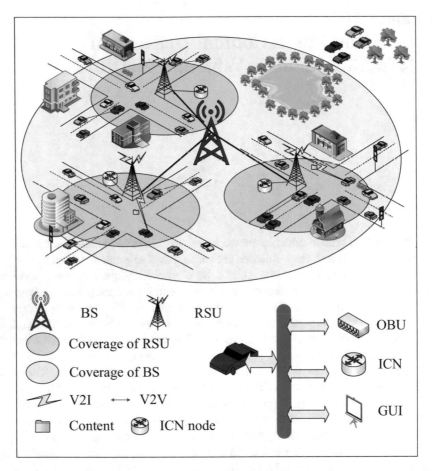

Fig. 2.1 Overview of the reputation based content delivery in information centric vehicular networks

Although the vehicular networks can provide vehicles with content delivery services, there come new challenges as follows.

- The amount of content requests in vehicular networks is not of the same order of magnitude as an increasing number of vehicles not only produce their own contents, but also request contents from other vehicles and infrastructures easier than before. Therefore, how to efficiently manage the content requests to improve the performance of vehicular networks becomes a challenge.
- With the high mobility of vehicles, the replicas of one content may be stored in different sites. As these replicas of the same content are managed according to their different locations in the current networks, the overhead to manage these replicas incurs huge operating costs. Consequently, how to design an efficient

approach to manage replicas of the same content in vehicular networks becomes an important issue.

To address these challenges, the information centric vehicular networks have advocated recently as a key enabling technology to manage the content requests of vehicles and control the replicas of the same content by integrating the information centric networking (ICN) and the vehicular networks [12–16]. With the adoption of ICN, content is delivered based on the interest instead of sending the conventional requested message. In this way, the vehicles which have the same interest can be managed efficiently, where a vehicle can obtain content from other vehicles that have the same interest instead of requesting the roadside infrastructures. In addition, unlike the conventional IP based content delivery, where the content is recognized by its IP address, content is delivered based on its naming information in the ICN [17–22]. As such, the replicas stored in different sites can be managed efficiently and the overhead to control these replicas can be reduced.

With the ever-increasing number of vehicles and the demands for quality of experience (QoE), when vehicles are benefited from the information centric vehicular networks, the problem of trust for content delivery, however, poses threats to the networks [23–28]. The reasons are as follows. First, unlike other existing networks, both the content forwarder and the content provider in the information centric vehicular networks are vehicles with high mobility. Subsequently, the connection via vehicle to vehicle (V2V) communication usually relies on the opportunity links and may have a long distance between the content requester and the content provider. Second, the ICN based content delivery does not provide the information about the content addresses and the information of the content provider. As a result, both content forwarder and content provider may not know the content requester very well, where they may have selfish and unfaithful behaviors. Therefore, the trust problem for content delivery in the information centric vehicular networks becomes a new challenge.

As an effort to tackle with this problem, in this chapter, a reputation based framework to deliver content in information centric vehicular networks is presented. In the networks, we first design a graphic user interface (GUI) for each vehicle to manage its interests and reputation values. Based on the reputation values of vehicles, both vehicles and infrastructures can select the suitable vehicles to deliver or relay contents. Then, the methods to calculate the vehicles' reputation and the cost paid for the content delivery services are detailed. By doing this, vehicles are encouraged to participate in the content delivery to increase their reputation and earn profits. After this, the update of vehicles' reputation is designed and a Bayes based scheme is used to find the untrustworthy vehicles in the networks. Simulation result proves that the proposed framework outperforms the existing scheme and is able to deliver contents more efficiently.

2.2   Overview of Information Centric Vehicular Networks

2.2.1   Content Delivery in Vehicular Networks

Recently, the vehicular networks are expected to provide contents to vehicles with high transmission rate, low end to end latency, low cost and consistent QoE. In vehicular networks, vehicles can not only request the needed contents from the infrastructures but also can upload the valuable contents to the infrastructures or help other vehicles and infrastructures to relay contents to earn profits. In this way, the content delivery mainly depends on two modes, i.e., V2V and V2I. To be specific, the V2V means that a vehicle can share its content directly with the vehicles in proximity. For example, vehicles in a congested area can construct an ad hoc network to share contents [29–34]. The other communication mode is V2I, where there are many infrastructures (i.e., RSUs and BSs) deployed along roads and each infrastructure can provide content delivery services to vehicles within its coverage. The RSUs typically have lower transmission power and smaller communication coverage compared to the BSs.

Although there are a large number of infrastructures deployed along roads, when a vehicle is not covered by any infrastructures, the vehicle needs to ask for the help of other vehicles for the content delivery. In addition, a vehicle usually cannot download the whole content from one infrastructure because the size of content keeps increasing in vehicular networks (e.g., high resolution video) and the contention among a large number of vehicles within the infrastructure's coverage. Therefore, the content relay plays an important role in the content delivery in the vehicular networks [35–38]. By using relay vehicles to forward the content, a vehicle can obtain the requested content more efficiently than that without relay vehicles no matter whether the vehicle is covered by an infrastructure or not.

2.2.2   ICN Based Content Delivery

To reduce the overhead of content management and the latency time to obtain content, ICN has proposed as a promising solution to facilitate the content delivery.

In the ICN, there are two types of packets, i.e., interest packet and data packet for the content delivery, where content is delivered based on the interest. In addition, content in the ICN is recognized and delivered based on its naming information rather than its IP address. Each ICN node consists of three parts.

- **Content store (CS):** The CS caches the replicas of content.
- **Pending interest table (PIT):** The PIT is used to cache the entries of the pending interests that do not cached in the CS.
- **Forwarding information base (FIB):** The FIB is used to record the paths of contents and forward interest to potential sources based on the cached paths.

For a vehicle which intends to obtain its interest content, the ICN based content delivery can be divided into two cases as follows.

Case 1: The Vehicle is in the Coverage of an Infrastructure
If a vehicle sends an interest to the infrastructure and the replica of this content is available in the CS of the ICN node associated with this infrastructure, this replica of the content will be provided to the vehicle directly. Otherwise, the ICN node will check whether there is a matching entry of the interest in the PIT. If there is a matching entry of the replica in the PIT, it means that the interest has already created an entry. If such an entry is not available in the PIT, the ICN node will check the FIB to wait for the requested content to be fetched from one of the other infrastructures by a suggested delivering path cached in the FIB.

Case 2: The Vehicle is Not in the Coverage of Any Infrastructures
If a vehicle is not in the coverage of any infrastructures and intends to obtain its interest content, the vehicle can broadcast the interest to other vehicles in its communication coverage. Then these vehicles check their ICN nodes for the requested content. If they can find the replica in the CS of one of these vehicles, the vehicle obtains the replica directly. Otherwise, if there is an entry which matches the interest in the PIT or the FIB of these vehicles' ICN nodes, the vehicle which requests the content only needs to wait for retrieving the content. On the contrary, if these vehicles cannot provide any information about the interest packet, the vehicle which requests the content needs to select a relay vehicle to forward its interest and let the relay vehicle do the same operations until one of the relay vehicles finds the replica of the content or finds the entry of the content either in PIT or FIB.

2.2.3 Challenges of Content Delivery in Information Centric Vehicular Networks

While the information centric vehicular networks can decrease the traffic load and improve the QoE of vehicles, the easy access mode and frequent content sharing among vehicles and infrastructures make the selfish and unfaithful behaviors of vehicles as a common phenomenon and more important than before. Although there are many efforts to study the incentive schemes to encourage vehicles to join in the relay process and pricing mechanisms to pay for the content providers [39–43], it cannot be aware of the untrustworthy vehicles which may obtain the payment but do nothing. These untrustworthy vehicles, however, bring new challenges to the content delivery in information centric vehicular networks as follows.

- As the caching and communication resources of both infrastructures and vehicles are not for free, a vehicle needs to pay for its requested content to compensate for the consumption of infrastructures and relay vehicles. However, an untrustworthy vehicle may obtain the payment without the contribution to relay or provide content to the content requester. Sometimes it even provides a wrong content or virus content. As a result, a vehicle may pay for the content delivery service more than one time until the vehicle obtains the content, where the cost of the vehicle is increased.

- Because of the mobility of vehicles, a requested content usually cannot be obtained instantly. To be specific, with the V2V mode, if a vehicle intends to download the requested content while the content is not available in the ICN nodes of the nearby vehicles, the vehicle needs to select a vehicle to relay its interest packet. In addition, if the content is not completely downloaded before the vehicle leaves the coverage of its connected infrastructure, it also needs other vehicles to relay the content. When the relay vehicle is untrustworthy, it may accept the payment and does not help the vehicle to obtain the content, resulting in long latency time.

- Different from other existing networks, the velocity of vehicles is higher than other mobile devices with the result that the interactions among vehicles and infrastructures typically spend a short time. Therefore, the QoE of vehicles mainly depends on the credit of the relay vehicles and the content providers. For example, an infrastructure delivers the content to a relay vehicle and lets the relay vehicle send the rest of the content to the content requester which is out of the infrastructure's coverage. If the relay vehicle is untrustworthy, it may only receive the content without delivering the content to the content requester. As a result, the QoE of the content requester may degrade significantly.

From the above analysis, it thus can be concluded that how to recommend honest vehicles to relay or provide content in information centric vehicular networks becomes a challenging issue.

2.3　Reputation Based Vehicular Networks

The reputation based vehicular networks has been advocated to meet the above challenges. Specially, each vehicle in the vehicular networks has its own reputation value, where the value is determined by the behaviors of the vehicle. If a vehicle fulfills the relay task in time or provides content to other vehicles successfully, it will obtain a higher reputation value. Otherwise, when a vehicle exhibits unfaithful behaviors, its reputation value will be decreased. The above challenges can be met because of the following reasons.

- **Decrease cost:** In the reputation based vehicular networks, if a vehicle can obtain content from other vehicles which are in its communication coverage, the vehicle then buys the content from a vehicle with a high reputation (higher than the average value) in the first time, where the extra cost paid for the content is avoided. On the other hand, if the vehicle cannot find the content provider within its communication range, the vehicle can send its request to an honest vehicle to forward the content request when the vehicle is out of any infrastructures. In addition, a discount according to the reputation value is enjoyed when the vehicle with a high reputation value buys content from infrastructures. In these ways, the cost of the vehicle for requesting its content can be decreased.

- **Decrease latency:** In the vehicular networks, vehicles can obtain content from infrastructures or other vehicles. When a vehicle is in the coverage of an infrastructure, it can download the content from the infrastructure directly. If the content is not completely downloaded when the vehicle leaves the coverage, the infrastructure will select an honest vehicle to relay the rest of the content, where the probability that the relay vehicle refuses to deliver or even drops the content is low. Similarly, when a vehicle requests a content from another vehicle, the latency will become lower when the relay vehicle and the content provider are honest than that they are dishonest. Therefore, the vehicle can obtain the content in time from other vehicles or infrastructures, where the latency can be decreased.
- **Consistent QoE improvement:** In the reputation based vehicular networks, when a vehicle intends to share content with others, the honest vehicles will be recommended. Meanwhile, the infrastructures in the networks select relay vehicles based on the reputation values. As such, the quality of the service is guaranteed. Specifically, the vehicle can obtain the requested content with low cost and latency, which leads to a high QoE to the vehicle. Moreover, the vehicle will not obtain wrong or virus content when sharing content with honest vehicles, where the QoE of this vehicle is increased. For example, when the content of a vehicle is not completely downloaded from an infrastructure yet the vehicle leaves the coverage, the infrastructure can select a forwarder which has a high reputation value to relay the content to keep a consistent QoE of the vehicle.

2.4 Framework of Reputation Based Content Delivery in Information Centric Vehicular Networks

In this section, we present our framework to address the challenges in the information centric vehicular networks. We first introduce the network architecture. Then, we design the framework of the reputation based content delivery in information centric vehicular networks. After that, the simulation is carried out to evaluate the performance of the proposed framework.

2.4.1 Network Architecture

The network architecture of the framework consists of on board units (OBUs), GUI, RSUs and BSs, as shown in Fig. 2.1.

- **OBUs:** In the information centric vehicular networks, vehicles can share content with each other by using OBUs. Each OBU is associated with an ICN node, where the ICN node has a limited caching capability to store contents in its CS

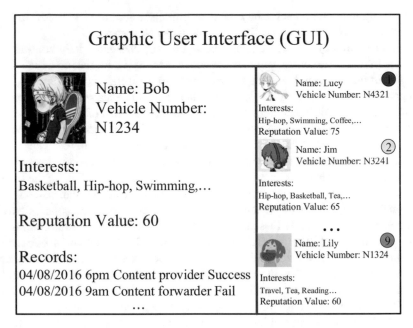

Fig. 2.2 Illustration of the GUI of a vehicle

selectively based on the vehicle's interests. Let C_i denote the capacity of the CS in the ICN of vehicle i. The content stored in vehicle i should satisfy

$$\sum_q s_q \leq C_i, \tag{2.1}$$

where s_q is the size of content q. The vehicle needs to pay for the content provided by other vehicles or infrastructures to compensate for the time, power and cache resources. Besides, vehicles can store valuable contents in advance and sell the contents requested by other vehicles or infrastructures to earn profits and increase reputation values.

- **GUI:** The GUI in a vehicle is used to post the social interests and the reputation value of this vehicle [44]. As shown in Fig. 2.2, by browsing the interfaces of other vehicles, a vehicle can make connection selectively based on the similarity of their interests and the reputation values of the nearby vehicles. Let $V_i = [v_{i,1}, \ldots, v_{i,m}, \ldots, v_{i,M}]$ denote the keywords vector of vehicle i's interests. For example, vehicle i sets its interests by $V_i = \{1, 0, 0\}$, with the three elements in sequence corresponding to popular music, basketball and swimming. This implies that vehicle i is interested in popular music, but dislikes basketball and swimming. Accordingly, the vector of the weight of the keywords is denoted by $W_i = [w_{i,1}, \ldots, w_{i,m}, \ldots, w_{i,M}]$. Then the interest matching between vehicles i and g can be evaluated by

$$sim(i, g) = \frac{\sum_m w_{i,m} w_{g,m}}{\sqrt{\sum_m w_{i,m}^2} \sqrt{\sum_m w_{g,m}^2}}. \tag{2.2}$$

Note that vehicles can set up and modify their interests by using the GUI. But they cannot change their reputation values which are managed and controlled by the background process of the system. If two vehicles have the similar interests and high reputations, they may share content via V2V communication. To facilitate the optimal vehicle selection, the GUI of each vehicle ranks the neighboring vehicles based on their reputation values and recommends them to the vehicle dynamically.

- **RSUs:** A group of RSUs is placed along the roads in a distributed manner in the information centric vehicular networks. ICN nodes are placed in the RSUs for information centric content delivery. For RSU j, the arrival rate of vehicles in its coverage is denoted as λ_j, which is based on the Poisson distribution [45–48]. All the RSUs are wired connect with the BS. Each RSU can provide vehicles with wireless connection when vehicles pass through its coverage. In addition, the GUI system of each vehicle is charged by RSUs to facilitate the regulation of vehicles' reputation values. The CS of each ICN node has a limited capacity and cannot store all replicas. Unlike the CS in vehicles which caches contents based on the interests of vehicles, the caching scheme in the CS of RSUs is based on the content size and content popularity to improve the utilization efficiency of the CS's caching resources.
- **BSs:** The BSs are mainly used to regulate the RSUs and process the information (such as the reputation values collected by different RSUs). On one hand, when a vehicle sends a content request to the nearby RSU, and there is an entry of the content in the FIB of this RSU, the RSU will fetch the content from its connected BS and deliver the content to the requested vehicle if the end of the path is the BS. On the other hand, based on the reputation values collected from vehicles, the BS can calculate the average reputation value in the information centric vehicular networks, where the average reputation value at time t is denoted as $A(t)$. Based on the reputation values of vehicles, the BS can find the untrustworthy vehicles in the networks.

2.4.2 Framework of Reputation Based Content Delivery in Information Centric Vehicular Networks

In this subsection, we show the framework of the reputation based content delivery in information centric vehicular networks. To be specific, we first introduce the process of the content delivery. Then, we detail the method to calculate the costs and reputation values of vehicles. After that, the reputation update scheme is designed.

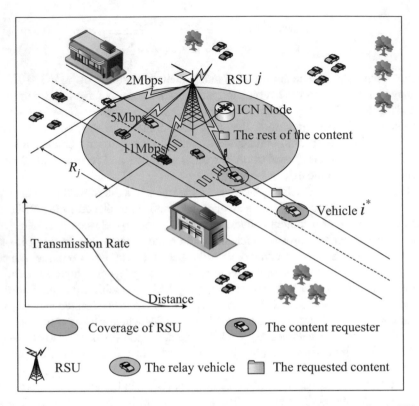

Fig. 2.3 Reputation based content delivery in information centric vehicular networks via V2I

- **Content delivery:** A vehicle can obtain content from an infrastructure or other vehicles. If vehicle i is in the coverage of RSU j, it can send its interest packet to download the content directly. If there is more than one vehicle in the coverage concurrently, vehicle i needs to compete with other vehicles to make the connection with the RSU. Caused by the fading channel, vehicle i will obtain a high transmission rate if the vehicle is close to RSU j, as shown in Fig. 2.3. Therefore, the RSU selects the vehicle to make the connection based on the distance and the reputation value, where the evaluation function is given by

$$i^* = \arg\max\{\delta_j \log\left(1 + \frac{r_i}{r_{\max}}\right) + (1 - \delta_j) \log\left(1 + \frac{d_{i,j}(t)}{R_j}\right)\}, i \in I'_j.$$
(2.3)

Here, I'_j is the set of vehicles in the coverage of RSU j. $r_{\max}$ is the maximum reputation value in this area. $d_{i,j}(t)$ is the distance between vehicle i and RSU j at time t. R_j is the coverage radius of RSU j. r_i is the reputation value of vehicle i. δ_j is the adjustment factor charged by RSU j. If the selected vehicle (i.e., i^*) cannot download the content completely and leaves the coverage, the RSU will

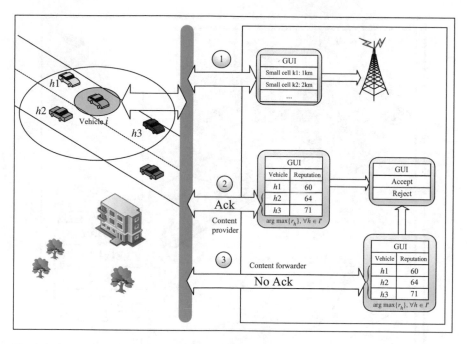

Fig. 2.4 Reputation based content delivery in information centric vehicular networks via V2V

select a vehicle with the highest reputation to relay the rest of the content to guarantee the QoE of vehicle i^*.

On the other hand, as shown in Fig. 2.4, when vehicle i is not in the coverage of any infrastructures and intends to obtain the content, the vehicle needs to make choices based on different cases. First, the GUI will recommend some locations of infrastructures which are deployed along the vehicle's trip. Second, the vehicle can explore V2V mode by broadcasting its interest packet to the vehicles within its communication coverage (e.g., $h1$, $h2$, and $h3$). When a vehicle (e.g., $h1$) has the data packet, the GUI will check the reputation value of this vehicle and make a decision, i.e., accept and buy the content if $r_{h1} \geq v_i A(t)$ or reject the content if $r_{h1} < v_i A(t)$. v_i is the adjustment factor charged by vehicle i. If more than one vehicles which cached the content are in the communication range, vehicle i will select the vehicle to obtain the requested content by

$$\arg\max\{r_h | r_h \geq v_i A(t)\}, h \in I'. \tag{2.4}$$

Here, I' is the set of vehicles which cache the content requested by vehicle i. If all the nearby vehicles do not cache the requested content, these vehicles then check their PIT to find an entry of the interest packet. If there is an entry that matches the interest, vehicle i then checks the vehicle's reputation and decides whether to retrieve the content from this vehicle or not. If there is no entry that can match

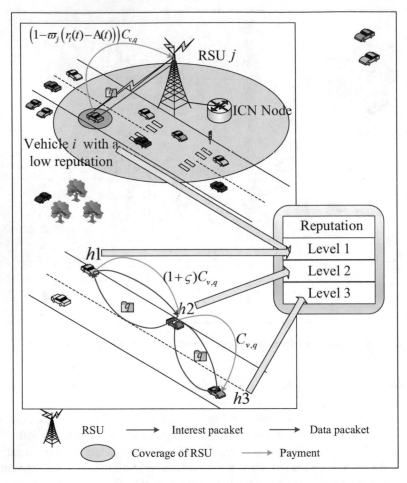

Fig. 2.5 Cost and reputation calculation of the reputation based content delivery in information centric vehicular networks

the interest, these vehicles then check their FIB. Similar to the PIT, if there is an entry that matches the interest, vehicle i then decides whether to retrieve the content by checking the reputation. On the contrary, if the requested content is not stored and there is no entry in PIT and FIB in these vehicles, vehicle i can select a relay vehicle from these vehicles to broadcast its interest packet based on their reputation values.

- **Cost and reputation calculation:** If the reputation value of vehicle i is low (lower than the average reputation value), it may not be selected by other vehicles to relay or provide content. As a result, the probability that the vehicle joins in the content delivery process as a content forwarder/provider becomes low. As shown in Fig. 2.5, if vehicle i needs to download content q from RSU j, $(1+\varpi_j(A(t)-r_i(t)))C_{s,q}$ will be paid, where $C_{s,q}$ is the original price of content

q if the content is obtained from infrastructures. ϖ_j is the adjustment factor to be charged by RSU j. Oppositely, when vehicle i has a high reputation (higher than the average reputation value), $(1 - \varpi_j(r_i(t) - A(t)))C_{s,q}$ will be paid for the content. If a vehicle (e.g., $h1$) obtains its requested content from another vehicle (e.g., $h3$), $(1 + n\varsigma)C_{v,q}$ will be paid for the service, where n is the number of relay vehicles which join in the content delivery process. In Fig. 2.5, we have $n = 1$. The payment for the relay vehicle is set to be $\varsigma C_{v,q}$, where $0 < \varsigma < 1$. $C_{v,q}$ is the payment for content q if the content is obtained from vehicles.

After a round of content delivery, the reputation values of the vehicles which join in the process will be calculated by the GUI. If the content is provided to the requested vehicle successfully, the reputation of the vehicles which participate in the process will be increased. However, when the content delivery is not fulfilled, the reputation of each participant will be decreased proportionally. For the requested content q, we divide different levels according to the importance of the participants in the content delivery process, as shown in Fig. 2.5. Specifically, the level of content requester is 1; the level of content forwarder is 2; and the level of content provider is 3. If the content is successfully provided, the reputation values of vehicles are changed by $r + r\Gamma_x, x \in \{1, 3\}$ for the content requester and content provider, respectively. For the content forwarder (e.g., h), the update of the reputation is based on the pending time for the content and its interests. If the content already has an entry in the PIT of vehicle h's ICN, the reputation of vehicle h is calculated by

$$r + r\Gamma_x + o\left(t_{h,q} - \frac{t'_q}{t_q}\right) + zsim(h, q), x \in \{2\}, \tag{2.5}$$

where o and z are adjustment coefficients. $sim(h, q)$ is the interest similarity of vehicle h's interests and the requested content q. If vehicle h is not interested in the content, we have $sim(h, q) = 0$. $t_{h,q}$ is the pending time of content q in the PIT of vehicle h's ICN. t_q is the maximum pending time of the content. t'_q is the pending time of content q before the new entry is created. On the other hand, if there is no entry in the PIT but an entry in the FIB, the vehicle adds a new entry of content q in its PIT. The reputation of vehicle h is calculated by

$$r + r\Gamma_x + o\left(\frac{t'_q}{t_q}\right) + zsim(h, q), x \in \{2\}. \tag{2.6}$$

By contrast, if the content delivery is not fulfilled, the reputation values of vehicles become $r(1 - \Gamma_x), x \in \{1, 2, 3\}$.

For a vehicle which has a low reputation value, it can obtain content from infrastructures with a high price (i.e., as the content requester) to increase its reputation value. To avoid the situation that vehicles which are friends increase their reputation with each other, when the times that a vehicle makes connections with another vehicle is larger than T_{max} in a period of time, the reputation values of the both will not be increased. Besides, the increased amount of vehicles' reputation value within a fixed period also has a threshold th_r.

- **Reputation update:** When the reputation value of a vehicle reaches the maximum value (i.e., r_{max}) in the information centric vehicular networks, its value will set to be r_{max}. It is noted that the reputation value of each vehicle is decreased over time in case of the vehicles which keep a high reputation and enjoy the discount from infrastructures while do not help other vehicles any more. We have

$$r(t) = re^{-\gamma t}, \tag{2.7}$$

where γ is the parameter of the attenuation factor. The reputation value of each vehicle is synchronized to an infrastructure by the GUI if the following conditions are satisfied: (1) the vehicle is in the coverage of an infrastructure; (2) the update interval of the reputation value between this time and the last time is larger than Δt.

Every $O \cdot \Delta t$, the BS collects the reputation of vehicles from the RSUs and divides vehicles into two groups, which are trustworthy and untrustworthy, respectively. As shown in Fig. 2.6, the BS evaluates a vehicle based on the times that the reputation of the vehicle is lower than the average reputation value and the times that the vehicle makes the connections with other vehicles, which are denoted as a_r and a_c, respectively. Then, the set of vehicle's features can be expressed by $\{a_r, a_c\}$. The set of categories is denoted by $y = \{0, 1\}$, where

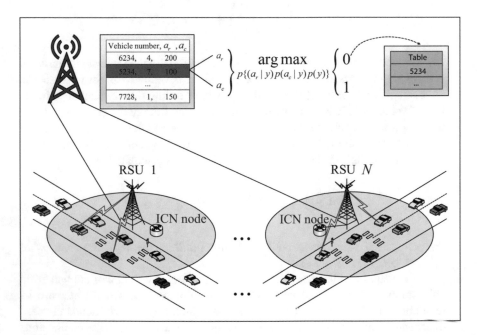

Fig. 2.6 The differentiate process of vehicles by BS

$y = 0$ means that the vehicle is untrustworthy and $y = 1$ otherwise. As both a_r and a_c are continuous variables, we discrete them as $a_r : \{a_r < A_r, a_r \geq 1 - A_r\}$ and $ac : \{a_c < A_c, a_c \geq 1 - A_c\}$. By doing this, the BS can divide vehicles into two groups by

$$y^* = \arg\max p(a_r|y) \cdot p(a_c|y) \cdot p(y), \tag{2.8}$$

where $p(y = 0)$, $p(y = 1)$, $p(a_r|y)$ and $p(a_c|y)$ can be obtained by using the historical data.

According to the value of y^*, a table is made by the BS, where the untrustworthy vehicles are blacklisted in this table. Then the BS transmits the table to all the RSUs. If a vehicle is listed in this table, its GUI will be closed and it will not be recommended to other vehicles in its coverage.

2.5 Simulation

In this section, we evaluate the successful transmission rate of vehicles in the coverage of an RSU by comparing the proposed reputation based framework with the conventional vehicular networks. We first introduce the network setting followed by the results analysis.

2.5.1 Setting

In the conventional vehicular networks, when a vehicle cannot download the whole content from the RSU, the RSU will randomly select a vehicle to relay the content. In the proposed reputation based content delivery framework, the RSU will select a trustworthy vehicle with the highest reputation value to forward the content. In the simulation, we divide the coverage of the RSU into 5 zones according to the fading channel, where the transmission rates are $[2, 5, 11, 5, 2]$ Mbps [49–51]. The arrival rate of vehicles entering the coverage of the RSU is determined by the Poisson distribution. The maximum values of vehicle density and velocity are 150 Veh/Km and 110 Km/h, respectively. The content size is uniformly distributed with a range of $[1, 4]$ MBytes. The vehicle density in the coverage of the RSU is selected from $\{20, 50\}$ Veh/Km. The communication range of the RSU and each vehicle are set to be 200 m and 100 m, respectively.

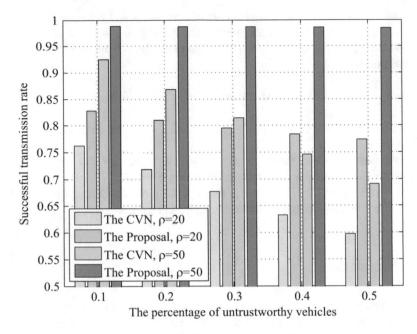

Fig. 2.7 Successful transmission rate with different percentages of untrustworthy vehicles in the area

2.5.2   Results Analysis

From the simulation result shown in Fig. 2.7, it can be seen that the proposal can obtain a higher successful transmission rate than the conventional vehicular networks with different percentages of untrustworthy vehicles in the networks. To be specific, the more untrustworthy vehicles in the coverage of the RSU are, the larger difference between the conventional vehicular networks and the proposal exists. In addition, with the increase of vehicles' density in the RSU's coverage, the number of vehicles to relay the content is increased with the result of a high rate of successful delivery. Therefore, the successful transmission rates in both the conventional vehicular networks and the proposal increase when the density of vehicles changes from 20 to 50 Veh/Km.

2.6   Summary

In this chapter, we have proposed a novel framework based on reputation to facilitate content delivery in information centric vehicular networks. By considering the behaviors of vehicles, we have designed a GUI for each vehicle to manage its interests and reputation value, where the reputation value is used to help both vehicles and infrastructures select the vehicles to join in the content delivery service. Then, we have designed the method to calculate the reputation of vehicles and the costs to pay for the content delivery service, respectively. With the incentives, vehicles are encouraged to take part in the content delivery to increase their reputation and earn profits. After this, we have developed a Bayes based scheme to find the untrustworthy vehicles in the networks. The simulation result has shown that the proposed framework can obtain a higher successful transmission rate with comparison to the existing scheme.

References

1. H. Zhou, N. Cheng, J. Wang, J. Chen, Q. Yu, X. Shen, Toward dynamic link utilization for efficient vehicular edge content distribution. IEEE Trans. Veh. Technol. **68**(9), 8301–8313 (2019)
2. C. Xu, Z. Zhou, Vehicular content delivery: a big data perspective. IEEE Wirel. Commun. **25**(1), 90–97 (2018)
3. Z. Su, Q. Xu, Y. Hui, M. Wen, S. Guo, A game theoretic approach to parked vehicle assisted content delivery in vehicular ad hoc networks. IEEE Trans. Veh. Technol. **66**(7), 6461–6474 (2017)
4. X. Wang, Z. Ning, X. Hu, L. Wang, B. Hu, J. Cheng, V.C.M. Leung, Optimizing content dissemination for real-time traffic management in large-scale internet of vehicle systems. IEEE Trans. Veh. Technol. **68**(2), 1093–1105 (2019)
5. F.A. Silva, A. Boukerche, T.R.M. Braga Silva, F. Benevenuto, L.B. Ruiz, A.A.F. Loureiro, Odcrep: origin–destination-based content replication for vehicular networks. IEEE Trans. Veh. Technol. **64**(12), 5563–5574 (2015)
6. Y. Hui, Z. Su, T.H. Luan, Collaborative content delivery in software-defined heterogeneous vehicular networks. IEEE/ACM Trans. Netw. **28**(2), 575–587 (2020)
7. A. Ghosh, V.V. Paranthaman, G. Mapp, O. Gemikonakli, J. Loo, Enabling seamless V2I communications: toward developing cooperative automotive applications in VANET systems. IEEE Commun. Mag. **53**(12), 80–86 (2015)
8. S. Pyun, W. Lee, D. Cho, Resource allocation for vehicle-to-infrastructure communication using directional transmission. IEEE Trans. Intell. Transp. Syst. **17**(4), 1183–1188 (2016)
9. V. Milanes, J. Villagra, J. Godoy, J. Simo, J. Perez, E. Onieva, An intelligent V2I-based traffic management system. IEEE Trans. Intell. Transp. Syst. **13**(1), 49–58 (2012)
10. M. Khabbaz, M. Hasna, C. M. Assi, A. Ghrayeb, Modeling and analysis of an infrastructure service request queue in multichannel V2I communications. IEEE Trans. Intell. Transp. Syst. **15**(3), 1155–1167 (2014)
11. V.A. Butakov, P. Ioannou, Personalized driver assistance for signalized intersections using V2I communication. IEEE Trans. Intell. Transp. Syst. **17**(7), 1910–1919 (2016)
12. C. Xu, W. Quan, H. Zhang, L.A. Grieco, Grims: green information-centric multimedia streaming framework in vehicular ad hoc networks. IEEE Trans. Circ. Syst. Video Technol. **28**(2), 483–498 (2018)
13. D. Grewe, M. Wagner, H. Frey, A domain-specific comparison of information-centric networking architectures for connected vehicles. IEEE Commun. Surv. Tutor. **20**(3), 2372–2388 (thirdquarter 2018)
14. S.H. Ahmed, S.H. Bouk, M.A. Yaqub, D. Kim, H. Song, J. Lloret, Codie: controlled data and interest evaluation in vehicular named data networks. IEEE Trans. Veh. Technol. **65**(6), 3954–3963 (2016)

15. Z. Su, Y. Hui, Q. Yang, The next generation vehicular networks: a content-centric framework. IEEE Wirel. Commun. **24**(1), 60–66 (2017)
16. V. Ortega, F. Bouchmal, J.F. Monserrat, Trusted 5G vehicular networks: blockchains and content-centric networking. IEEE Vehic. Technol. Mag. **13**(2), 121–127 (2018)
17. C. Fang, H. Yao, Z. Wang, W. Wu, X. Jin, F.R. Yu, A survey of mobile information-centric networking: Research issues and challenges. IEEE Commun. Surv. Tutor. **20**(3), 2353–2371 (2018)
18. M.F. Al-Naday, M.J. Reed, D. Trossen, K. Yang, Information resilience: source recovery in an information-centric network. IEEE Netw. **28**(3), 36–42 (2014)
19. D. Grewe, M. Wagner, H. Frey, A domain-specific comparison of information-centric networking architectures for connected vehicles. IEEE Commun. Surv. Tutor. **20**(3), 2372–2388 (2018)
20. C. Fang, F.R. Yu, T. Huang, J. Liu, Y. Liu, A survey of green information-centric networking: research issues and challenges. IEEE Commun. Surv. Tutor. **17**(3), 1455–1472 (2015)
21. K. Zheng, Y. Cui, X. Liu, X. Wang, X. Jiang, J. Tian, Asymptotic analysis of inhomogeneous information-centric wireless networks with infrastructure support. IEEE Trans. Veh. Technol. **67**(6), 5245–5259 (2018)
22. M. Zhang, H. Luo, H. Zhang, A survey of caching mechanisms in information-centric networking. IEEE Commun. Surv. Tutor. **17**(3), 1473–1499 (2015)
23. H. Xia, S. Zhang, Y. Li, Z. Pan, X. Peng, X. Cheng, An attack-resistant trust inference model for securing routing in vehicular ad hoc networks. IEEE Trans. Veh. Technol. **68**(7), 7108–7120 (2019)
24. K. Rostamzadeh, H. Nicanfar, N. Torabi, S. Gopalakrishnan, V.C.M. Leung, A context-aware trust-based information dissemination framework for vehicular networks. IEEE Internet Things J. **2**(2), 121–132 (2015)
25. Z. Yang, K. Yang, L. Lei, K. Zheng, V.C.M. Leung, Blockchain-based decentralized trust management in vehicular networks. IEEE Internet Things J. **6**(2), 1495–1505 (2019)
26. W. Li, H. Song, Art: an attack-resistant trust management scheme for securing vehicular ad hoc networks. IEEE Trans. Intell. Transport. Syst. **17**(4), 960–969 (2016)
27. S. Ahmed, S. Al-Rubeaai, K. Tepe, Novel trust framework for vehicular networks. IEEE Trans. Veh. Technol. **66**(10), 9498–9511 (2017)
28. H. Hu, R. Lu, Z. Zhang, J. Shao, Replace: a reliable trust-based platoon service recommendation scheme in VANET. IEEE Trans. Veh. Technol. **66**(2), 1786–1797 (2017)
29. Y. Mylonas, M. Lestas, A. Pitsillides, P. Ioannou, V. Papadopoulou, Speed adaptive probabilistic flooding for vehicular ad hoc networks. IEEE Trans. Veh. Technol. **64**(5), 1973–1990 (2015)
30. S. Bitam, A. Mellouk, S. Zeadally, Bio-inspired routing algorithms survey for vehicular ad hoc networks. IEEE Commun. Surv. Tutor. **17**(2), 843–867 (2015)
31. T. Qiu, X. Wang, C. Chen, M. Atiquzzaman, L. Liu, TMED: a spider-web-like transmission mechanism for emergency data in vehicular ad hoc networks. IEEE Trans. Veh. Technol. **67**(9), 8682–8694 (2018)
32. D. He, S. Zeadally, B. Xu, X. Huang, An efficient identity-based conditional privacy-preserving authentication scheme for vehicular ad hoc networks. IEEE Trans. Inf. Foren. Sec. **10**(12), 2681–2691 (2015)
33. K. Golestan, B. Khaleghi, F. Karray, M.S. Kamel, Attention assist: a high-level information fusion framework for situation and threat assessment in vehicular ad hoc networks. IEEE Trans. Intell. Transport. Syst. **17**(5), 1271–1285 (2016)
34. X. Huang, J. Wu, W. Li, Z. Zhang, F. Zhu, M. Wu, Historical spectrum sensing data mining for cognitive radio enabled vehicular ad-hoc networks. IEEE Trans. Depend. Secure Comput. **131**, 59–70 (2016)
35. M.F. Feteiha, H.S. Hassanein, Enabling cooperative relaying VANET clouds over LTE-A networks. IEEE Trans. Veh. Technol. **64**(4), 1468–1479 (2015)
36. D. Tian, J. Zhou, Z. Sheng, M. Chen, Q. Ni, V.C.M. Leung, Self-organized relay selection for cooperative transmission in vehicular ad-hoc networks. IEEE Trans. Veh. Technol. **66**(10), 9534–9549 (2017)

37. D. Wang, P. Ren, Q. Du, L. Sun, Y. Wang, Security provisioning for miso vehicular relay networks via cooperative jamming and signal superposition. IEEE Trans. Veh. Technol. **66**(12), 10732–10747 (2017)
38. P. Gu, C. Hua, R. Khatoun, Y. Wu, A. Serhrouchni, Cooperative antijamming relaying for control channel jamming in vehicular networks. IEEE Trans. Veh. Technol. **67**(8), 7033–7046 (2018)
39. C. Lai, K. Zhang, N. Cheng, H. Li, X. Shen, SIRC: a secure incentive scheme for reliable cooperative downloading in highway VANETs. IEEE Trans. Intell. Transp. Syst. **18**(6), 1559–1574 (2017)
40. Y. Hui, Z. Su, T. H. Luan, J. Cai, Content in motion: an edge computing based relay scheme for content dissemination in urban vehicular networks. IEEE Trans. Intell. Transp. Syst. **20**(8), 3115–3128 (2019)
41. F. Tseng, Y. Liu, J. Hwu, R. Chen, A secure reed–solomon code incentive scheme for commercial ad dissemination over VANETs. IEEE Trans. Veh. Technol. **60**(9), 4598–4608 (2011)
42. T. Wang, L. Song, Z. Han, B. Jiao, Dynamic popular content distribution in vehicular networks using coalition formation games. IEEE J. Sel. Area. Commun. **31**(9), 538–547 (2013)
43. Y. Hui, Z. Su, S. Guo, Utility based data computing scheme to provide sensing service in internet of things. IEEE Trans. Emerging Topics in Computing **7**(2), 337–348, Apr. 2019.
44. T.H. Luan, R. Lu, X. Shen, F. Bai, Social on the road: enabling secure and efficient social networking on highways. IEEE Wirel. Commun. **22**(1), 44–51 (2015)
45. M.H. Cheung, F. Hou, V.W.S. Wong, J. Huang, DORA: dynamic optimal random access for vehicle-to-roadside communications. IEEE J. Sel. Areas Commun. **30**(4), 792–803 (2012)
46. H. Zhou, B. Liu, F. Hou, T.H. Luan, N. Zhang, L. Gui, Q. Yu, X.S. Shen, Spatial coordinated medium sharing: optimal access control management in drive-thru internet. IEEE Trans. Intell. Transport. Syst. **16**(5), 2673–2686 (2015)
47. Z. Su, Y. Hui, T.H. Luan, S. Guo, Engineering a game theoretic access for urban vehicular networks. IEEE Trans. Veh. Technol. **66**(6), 4602–4615 (2017)
48. Y. Zhuang, J. Pan, V. Viswanathan, L. Cai, On the uplink mac performance of a drive-thru internet. IEEE Trans. Veh. Technol. **61**(4), 1925–1935 (2012)
49. T.H. Luan, X. Ling, X. Shen, Mac in motion: impact of mobility on the mac of drive-thru internet. IEEE Trans. Mobile Comput. **11**(2), 305–319 (2012)
50. M. Xing, J. He, L. Cai, Maximum-utility scheduling for multimedia transmission in drive-thru internet. IEEE Trans. Veh. Technol. **65**(4), 2649–2658 (2016)
51. W.L. Tan, W.C. Lau, O. Yue, T.H. Hui, Analytical models and performance evaluation of drive-thru internet systems. IEEE J. Sele Areas Commun. **29**(1), 207–222 (2011)

Chapter 3
Contract Based Edge Caching in Vehicular Networks

Due to the massive requirements of various kinds of services requested by the expanding scale of vehicles, current caching schemes in vehicular networks face the challenges to decrease the latency, increase the efficiency of the networks and improve the quality of experience (QoE) of vehicles. As an effort, in this chapter, a novel contract based edge caching framework by considering the traffic status in vehicular networks is presented. To be specific, by jointly analyzing the traffic status and the requirements of vehicles, the contracts can be signed between edge caching devices (ECDs) and social vehicles in advance as the prior knowledge to facilitate the edge caching in vehicular networks. Then, based on the social relationships among vehicles, we present the detailed process of contract based edge caching in vehicular networks including content caching, content replacement and content delivery among social vehicles. For each ECD, content caching and replacement are based on three factors which are the contracts signed between vehicles and ECDs, the popularity of contents and the relevance among contents. As for the content delivery among social vehicles, we consider two different cases according to their social ties. Simulation experiment is carried out to demonstrate the performance of the proposed framework and the result shows that our proposed framework can outperform the conventional schemes in terms of the average transmission delay.

3.1 Introduction

With the rapid advance of mobile communication and intelligent transportation technologies, vehicular networks are expected to provide vehicles with not only pleasant and safety driving but also various kinds of services such as multimedia entertainment and social interactions on the road [1–5]. With a device called by on board units (OBUs), the vehicles in the existing vehicular networks can cache contents and share the contents with other vehicles through vehicle to vehicle (V2V)

© Springer Nature Switzerland AG 2021
Z. Su et al., *The Next Generation Vehicular Networks, Modeling, Algorithm, and Applications*, Wireless Networks, https://doi.org/10.1007/978-3-030-56827-6_3

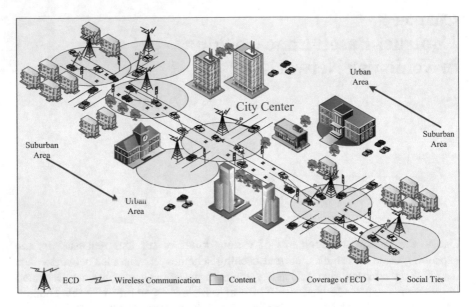

Fig. 3.1 Overview of the edge caching in vehicular networks

communications [6–10]. To satisfy the increasing number of demands requested by vehicles and enhance the contents sharing among social vehicles, the design of a novel framework in vehicular networks becomes an important issue.

Edge caching has been advocated to facilitate the content dissemination and alleviate the traffic load in vehicular networks recently [11–14]. As shown in Fig. 3.1, by deploying light-weight cloud-like edge caching devices (ECDs) along the roadside in vehicular networks, vehicles can download contents from the nearby ECDs directly. To be specific, each ECD consists of a content store to cache contents, an operation device to compute data and a transceiver to make the connection with other devices (i.e., cloud server and social vehicles). With edge caching, social vehicles are able to connect to ECDs by vehicle to infrastructure (V2I) communications instead of obtaining contents from the remote cloud server if the requested contents are cached in the content store of ECDs. Consequently, ECDs can provide convenient services to social vehicles at the edge of the networks.

There are some advantages by using ECDs to serve social vehicles in vehicular networks, which are detailed as follows [15–18].

- As compared to cloud caching, ECDs can provide services with the increased data transmission rate and decreased service latency. As a result, the latency to retrieve the content can be reduced.
- Download content from ECDs avoids the back-and-forth traffic between the cloud server and social vehicles. In this way, both the traffic load of the backbone networks and energy consumption can be reduced. Therefore, the efficiency of the networks is enhanced.

- Contents are cached in the ECDs with a distributed manner in vehicular networks. On the one hand, ECDs can work together with social vehicles to cache content to enhance the content sharing among social vehicles. On the other hand, ECDs are charged by the cloud server to update the content cached in their content stores. As such, social vehicles can obtain and update the content in time, where the QoE of vehicles can be improved.

While social vehicles can benefit from edge caching, the performances including latency to obtain content, efficiency of the networks and QoE of vehicles still need to be further studied. This is mainly attributed to the following reasons.

- It is costly to deploy ECDs everywhere to serve social vehicles.
- The ECDs placed in the vehicular networks are usually constrained by cache capacity and transmission coverage.
- The number of requests, which are generated by the ever-increasing scale of connected social vehicles, grows rapidly.

As a result, new challenges of the edge caching in vehicular networks come as follows.

- **Long latency of content transmission:** When the requested content is not cached in the nearby ECD, the ECD needs to connect the cloud server to fetch the content. Then the content is delivered to the vehicle after the ECD downloads the content from the cloud server, which results in a long transmission latency.
- **Inefficient content sharing among social vehicles:** Caused by the limited cache capacity of both vehicles and ECDs, it is impossible to store all contents at the edge. On the other hand, the contents in vehicular networks change over time, when purely relying on the content shared among social vehicles, the contents may not be updated in time. Accordingly, the networks have low efficiency.
- **Low QoE of vehicles:** When the efficiency of the vehicular networks is low, the frequency of content sharing among vehicles also becomes low, which leads to a long latency to obtain the requested content and a low QoE for vehicles.

To address the above problems, in this chapter, we present a novel framework for edge caching in vehicular networks by taking the traffic status into account. As shown in Fig. 3.1, a group of ECDs are distributed in the networks to store the replicas of vehicular contents. The cloud server has the responsibility for managing and coordinating ECDs. By analyzing the traffic status and the requirements of vehicles, the contracts can be signed between ECDs and social vehicles in advance. Then, we present the detailed process of contract based edge caching in vehicular networks including content caching, content replacement and content delivery among social vehicles. The design of content caching and replacement in ECDs is based on the contracts signed between social vehicles and ECDs, the popularity of contents and the relevance among contents. As for the content delivery among social vehicles, we consider two different cases according to their social ties. Finally, the simulation experiment is carried out to evaluate the efficiency of the proposed framework.

3.2   Edge Caching in Vehicular Social Networks

3.2.1   Vehicular Social Networks

Social networks, such as mobile social networks [19–21] and online social networks [22–24], have attracted much attention in recent years. In the social networks, social ties are constructed among users through sharing contents and applications. For example, two users may become friends if they are interested in the same content. In general, users are more likely to communicate with their social friends than making connection with strangers. By forming the social ties among users, they can find potential friends who have the same interests to further improve the content sharing in the social networks.

Similarly, due to location proximity, vehicles on the roads tend to be interested in similar information, such as road traffic and weather conditions [25]. In addition, vehicles in the vehicular networks may have the specific characteristics, such as traveling with the same destination and parking at the same gas station or parking lot. These features enable them to have common interests and share contents with each other via direct V2V communication. In this way, strange vehicles may construct social ties through the social communications [26–31]. For example, two vehicles which have similar travel routes may become friends after more than one time encountering and sharing contents.

In addition to the similar points, vehicles in vehicular networks also have unique features.

- The frequency of content sharing in vehicular networks is low and the communication duration is short due to the high mobility of vehicles. To guarantee content delivery, a large number of vehicles which can provide contents are needed.
- Unlike online social networks and mobile social networks, where connected social users typically already know each other in the real-world or are closely associated in a static mode, vehicles on roads are anonymous and unknown to each other.
- Different from other devices, vehicles in the vehicular networks are usually capacity constrained and they often cache contents based on their own interests. As a result, the efficiency of contents sharing among vehicles has been significantly decreased.

3.2.2   Edge Caching in Vehicular Networks

With an ever-increasing scale of intelligent transportation, massive content requests generated by vehicles lead to a heavy network load. The ECDs deployed at the infrastructures in vehicular networks to provide vehicles with edge caching services have been proposed as an efficient solution to alleviate the data traffic and facilitate the content delivery. The motivation of edge caching is by placing light-weight cloud-like ECDs at the proximity of vehicles so that ECDs can provide the contents and application services as close as possible to vehicles. In this way, ECDs can serve vehicles with a direct connection instead of connecting with the cloud server. More importantly, as deployed at localized sites, ECDs can provide customized services that are more desirable to vehicles.

To satisfy the increasing number of demands generated by vehicles, the ECDs can pre-cache the requested contents in their content store. After a vehicle downloads the requested content from an ECD, the content then can be shared among vehicles to improve the frequency of content exchange in vehicular networks. By using edge caching in vehicular networks, on the one hand, vehicles can download the contents pre-cached in the ECDs when they are in the coverage to reduce the transmission delay. On the other hand, by providing various contents to vehicles, the contents which have been cached in the content store of vehicles can be updated in time to keep the pace with the change of the contents cached in ECDs.

As the cache capacity of each ECD is limited, it is impossible to cache all the contents in the networks. Therefore, the ECDs need to cache contents selectively to improve the content delivery efficiency. In fact, the ECDs placed in the networks typically have different locations, where the traffic status and popularity of contents in different locations are usually different. Therefore, with different edge computing strategies, the content delivery by using ECDs in the vehicular networks may have different performances.

3.2.3 Challenges of Edge Caching in Vehicular Networks

Although edge caching can enrich the content delivery in the vehicular networks, the research challenges come as follows.

- **Latency of content transmission:** A requested content, especially the multimedia content, usually cannot be obtained instantly when a vehicle downloads the content from an ECD. When the vehicle is in the coverage of the ECD, on the one hand, the time when the vehicle stays in the coverage is short. On the other hand, the probability of collision to connect with the ECD increases rapidly with the large number of vehicles in the coverage. If the vehicle is selected by the ECD to make the connection, when the content is cached in the ECD, the ECD delivers the content to the vehicle directly. Otherwise, the ECD needs to connect the cloud server to fetch the content and then delivers it to the vehicle, which results in a long transmission latency. In addition, most of the existing caching methods of ECDs are based on the popularity of contents. The cooperation among ECDs to cache contents is ignored. As a result, different ECDs may cache the same content, which also leads to a long latency to provide contents to social vehicles. Therefore, how to reduce the latency of content transmission in vehicular networks by considering the social ties among vehicles becomes a new issue.

- **Efficiency of networks:** Due to the limited capacity of storage, the efficiency of the networks is therefore degraded if the storage capacity of each ECD is not fully utilized to cache contents. On the other hand, caused by the limited caching capacity of the content store equipped in vehicles, vehicles may give a high priority to cache the contents which they have interest in. As such, the content sharing among vehicles has the constraint that there is no content sharing if no vehicles have the similar interest. When there is no content exchange, the vehicular networks will fail to be constructed. Thus, how to enhance the efficiency of the edge caching in the vehicular networks becomes a challenging issue.

- **QoE of vehicles:** In the vehicular networks, the QoE of vehicles mainly depends on the experience to share content with others and the latency to request content from the nearby ECDs. With the V2V mode, a social vehicle first broadcasts a request, then other vehicles within its communication range will check if they have cached the requested content. If the requested content is not cached, the vehicle needs to wait to enter the coverage of an ECD to download the content. In addition, the cached content in vehicles may not be updated in time due to the sparse distribution of ECDs. As such, vehicles often cannot obtain the latest version of the requested content. On the other hand, when the efficiency of the vehicular networks is low, the frequency of content sharing among vehicles also becomes low, which leads to a low experience for vehicles. Therefore, with the edge caching, how to improve the QoE of social vehicles in the vehicular networks is a non-trivial issue.

3.3 Contract Based Edge Caching in Vehicular Networks

The edge caching based on the contracts signed by social vehicles and ECDs by considering the traffic status is a promising solution to meet the above challenges. The traffic status, in general, has some unique features. For example, most vehicles drive from suburban to urban area for working, traveling, shopping, etc., in the morning. In contrast to this, the direction of most vehicles becomes urban area to suburban in the evening. As such, the ECDs in the suburban area may be busier than those in the urban area in the morning. By considering the above characteristics for edge caching in vehicular networks, vehicles can sign contracts with the ECDs to obtain contents. In this way, the ECDs can manage and pre-cache the requested contents efficiently. In addition, each ECD can collaboratively cache contents with its neighbor ECDs after analyzing the relevance of the contents. The above challenges of edge caching can be addressed because of the following reasons.

- **Reduce the latency of content transmission:** The caching among ECDs can efficiently avoid the content redundancy in the vehicular networks. With the contract based edge caching, a vehicle in the vehicular networks has three ways to obtain the requested content. First, the vehicle signs a contract with the ECD in advance, and the ECD pre-caches the content. When the vehicle goes through the coverage of the ECD, the content will be provided to this vehicle directly, where the latency to fetch the content from the cloud server is avoided. Second, the vehicle does not sign a contract with the ECD and connect to the ECD to request the content. If the content is cached in the ECD according to the content size and popularity, the content will also be provided to the vehicle. Otherwise, the ECD resorts to the cloud server to provide the content. Third, the vehicle can obtain the content from its social friends. With more frequent content sharing in the vehicular networks, vehicles are more likely to be friends with each other. As a result, the probability that the vehicle encounters a vehicle which has cached the requested content becomes high, where the latency to obtain the content is decreased.

- **Enhance the efficiency of networks:** The content cached in each ECD jointly considers the popularity of contents and the requirements of vehicles, where the utilization of the content store in each ECD can be improved. Based on the contracts and the collaboration among ECDs, the contract based edge caching in the vehicular networks can provide the requested content to vehicles more conveniently, avoiding the back-and-forth traffic between the cloud server and social vehicles. This not only saves the backbone bandwidth, but also significantly motivates the content sharing among vehicles. With more frequent content exchange in the vehicular networks, the contract based edge caching therefore represents a promising solution to enhance the efficiency of the vehicular networks.

- **Improve the QoE of vehicles:** Unlike the traditional caching methods, such as caching contents based on the popularity of them, the contract based edge caching leverages the potential relations in the contents. By taking into account the traffic status and the requirements of vehicles, the content distribution in the networks is in a collaborative manner. As such, vehicles can have more frequent content sharing and the latency to download content can be decreased, where the QoE of vehicles is increased. More importantly, with the coordination by the cloud server and the cooperation among ECDs, contents can be updated in both the ECDs and vehicles in time, with the result that the quality of services is guaranteed.

3.4   Framework of Contract Based Edge Caching in Vehicular Networks

In this section, we detail the proposed contract based framework for edge caching in vehicular networks. As shown in Fig. 3.2, the framework consists of cloud server, edge caching devices and social vehicles.

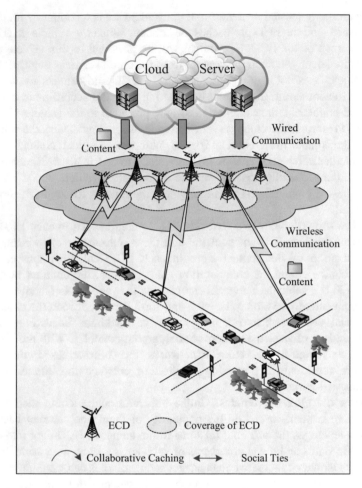

Fig. 3.2 Architecture of the edge caching in vehicular networks

3.4.1   Network Architecture

- **Cloud server:** The cloud server in the vehicular networks is made up of a large number of servers for enabling ubiquitous and convenient network access to provide applications and services to social vehicles [32–36]. All the contents in the networks are cached in the cloud server and the dynamic content update and the change of the popularity of each content are also charged by the cloud server. The cloud server connects with the ECDs through wired communication to distribute contents to the ECDs at the edge. On the other hand, the cloud server has the responsibility for charging and managing all the ECDs. The cloud server collects the traffic information and the contracts signed between social vehicles and ECDs and then controls and coordinates the content distribution in the networks by allocating different contracts to ECDs at different time.

- **Social vehicles:** Each social vehicle in the vehicular networks is equipped with an OBU which consists of a content store and a content transceiver. The content store of an OBU is used to cache the contents which are interested by the social vehicle. Different OBUs can communicate with each other through wireless communication. In this way, vehicular contents can be shared and exchanged among social vehicles in the vehicular networks. On the other hand, each OBU can communicate with the ECDs deployed along the roadside when the social vehicle goes through their communication coverage. When a social vehicle intends to obtain a content, it can send the request to other vehicles with social ties or download the content from the nearby ECD when the vehicle enters the coverage of the ECD.
- **Edge caching devices:** A group of ECDs is placed along the roadside in a distributed manner. These ECDs are connected with the cloud server so that they can leverage the rich caching resources of the cloud server to serve vehicles at the edge. To reduce the back-and-forth traffic and latency between the cloud server and social vehicles, ECDs can selectively pre-cache a part of the contents in their content stores. If a requested content is available in the content store of the ECD, it will be provided to the vehicle. Otherwise, the ECD delivers the content to the vehicle after connecting the cloud server to fetch the content. Each ECD can provide vehicles with wireless connections when vehicles pass through its coverage area. Caused by the fading channel, when a vehicle communicates with an ECD, the coverage area of the ECD can be divided into several zones [37–41]. The set of zones is denoted by $Z = \{1, \ldots, z, \ldots, Z\}$. The transmission rate when the vehicle downloads its requested content changes with the distance between the vehicle and the ECD.

3.4.2 Framework of Contract Based Edge Caching in Vehicular Networks

The traffic status in vehicular networks has its unique features, as we can see from Fig. 3.1, vehicles drive from the suburban to the urban area in the morning. Therefore, the edge caching in vehicular networks needs a novel framework which is different from the conventional methods by considering the traffic status to provide contents to vehicles efficiently. In this subsection, we detail the mechanism of the contract based edge caching framework in the vehicular networks. To be specific, we first introduce the contracts signed between vehicles and ECDs. After this, we show the contract based collaborative content caching in the content store of ECDs. Next, the contract based collaborative content replacement in ECDs is designed. Finally, the content delivery among social vehicles in our framework is introduced.

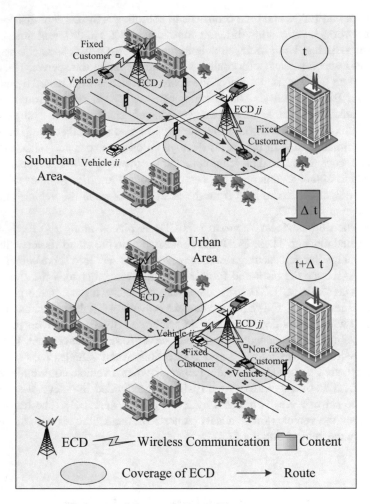

Fig. 3.3 Content provided with different types of vehicles

- **Contract signed between social vehicles and ECDs:** As the ECDs placed in the area have fixed locations and each vehicle has unique features, such as license plate number, vehicles can sign long-term contracts with ECDs in advance to obtain contents. As shown in Fig. 3.3, vehicle i ($i \in \{1, \ldots, i, \ldots, I\}$), which is interested in content q, drives from suburban to the urban area every morning. Then the vehicle can sign a contract with the nearest ECD (e.g., j) every week/month/year. As such, vehicle i becomes a fixed customer for ECD j. At time t, when vehicle i enters the coverage of ECD j, content q will be delivered to vehicle i directly. After a period of time, i.e., Δt, vehicle i enters the coverage of ECD jj, if vehicle i has not signed a contract with this ECD before, vehicle i is the non-fixed customer. By contrast, vehicle ii in the coverage of ECD jj is the fixed customer which has signed a contract with ECD jj. The information of a contract consists of the following parts: the target vehicle, the name of the content and the time range to provide the content. For example, the contract of vehicle i is $\{i, \text{content } q, t_{\min}, t_{\max}\}$. If the time to download the content from the ECD is over, the ECD has the right to cancel the contract and free up the storage capacity.

- **Contract based content caching in ECDs:** As the content store in each ECD has a limited storage capacity, how to determine the priority to cache contents becomes important. Let C be the capacity of the content store in each ECD. For ECD j, we divide C into two parts which are denoted as $\gamma_j \cdot C$ and $(1 - \gamma_j) \cdot C$, respectively. The $\gamma_j \cdot C$ is used to cache the contract based contents which are requested by fixed customers, as shown in Fig. 3.4. Note that variable γ_j changes over time based on the number of contracts signed by vehicles and the time to provide these contents. The second part is the popularity based caching. Except the contents which have been cached in the first part, the remaining contents are stored in the second part based on the popularity and the size of contents. As the request of contents follows a Zipf-like distribution [42–47], the probability that a request is for content q can be computed by

$$\frac{\left(\sum_q \frac{1}{q^z}\right)^{-1}}{r_q{}^z}, \tag{3.1}$$

where z is the Zipf parameter and r_q is the ranking of content q. The priority of content q in the second part of the ECD's content store can be determined by

$$\lambda \frac{\left(\sum_q \frac{1}{q^z}\right)^{-1}}{r_q{}^z} + (1 - \lambda) \frac{s_{\max} - s_q}{s_{\max}}, \tag{3.2}$$

where s_q and $s_{\max}$ are the size of content q and the maximum content size in the networks. λ is the parameter to balance the popularity of the content and the content size.

By considering the traffic status that the vehicle drives from suburban to the urban area in the morning, the collaborative caching among ECDs is studied in our framework. Let $V_q = [v_{q,1}, \ldots, v_{q,m}, \ldots, v_{q,M}]$ denote the keywords vector of content q. Accordingly, the vector of the weight of the keywords is denoted by $W_q = [w_{q,1}, \ldots, w_{q,m}, \ldots, w_{q,M}]$. Then the relevance between content q and content q' can be evaluated by

$$sim(q, q') = \frac{\sum_m w_{q,m}}{\sqrt{\sum_m w_{q,m}^2} \sqrt{\sum_m w_{q',m}^2}}. \tag{3.3}$$

Let A and B denote the event that vehicle i has downloaded content q and the event that vehicle i intends to download content q', respectively. Then the

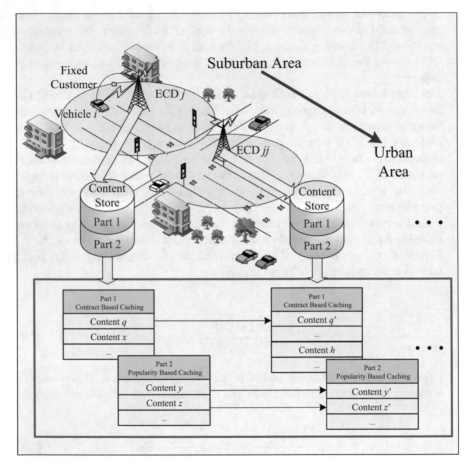

Fig. 3.4 Content caching in ECDs

probability that vehicle i intends to download content q' after obtaining content q can be denoted by $P(B|A)$. By balancing the above two factors, the relevance between the two contents can be expressed as

$$\Gamma \cdot sim(q, q') + (1 - \Gamma)P(B|A), \tag{3.4}$$

where Γ is the balance factor.

Compared with the ECD which has the farthest distance to the city center, on one hand, other ECDs need to cache their contracts assigned by the cloud server. On the other hand, the contents which have the highest relevance with the contents cached in the former ECD also need to be cached in the ECD's content store. As shown in Fig. 3.4, the contents cached in ECD j include the contracts signed by vehicles and the popularity based caching. In comparison, the content

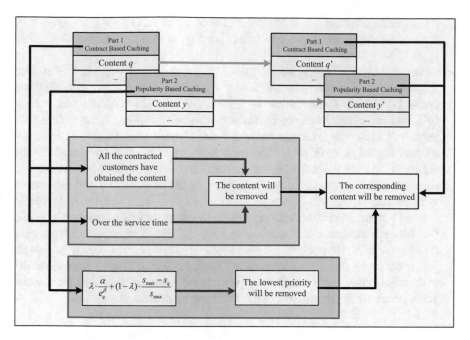

Fig. 3.5 Content replacement in ECDs

store of ECD jj not only caches the contents signed by vehicles, but also caches the content which has high relevance with the content cached in ECD j.

- **Contract based content replacement in ECDs:** Every a fixed period of time, the cloud server updates the contents cached in the content store of ECDs. The content replacement of ECD j is different based on different parts of its content store. As shown in Fig. 3.5, when it needs room for new contents to cache in the first part of the content store, the content which satisfies one of the following conditions will be removed: (1) the content has been provided to all the fixed customers; (2) the service deadline to obtain the content is exceeded. Otherwise, if both the above two conditions are not satisfied, the content which has the lowest priority cached in the second part will be replaced by the new content. When the second part of the content store is full and needs room for new contents, the stored content with the lowest priority will be replaced. Namely, the content removed from the content store of the ECD is selected by

$$\arg\min\left\{\lambda\frac{\left(\sum_q \frac{1}{q^z}\right)^{-1}}{r_q{}^z} + (1-\lambda)\frac{s_{\max} - s_q}{s_{\max}}\right\}. \tag{3.5}$$

As for ECD jj, the content replacement is based on the content update in ECD j. Specifically, when the content in the content store of ECD j is removed, the corresponding content cached in the content store of ECD jj is also replaced by the content which has the highest relevance degree with the new content cached in ECD j.

- **Content delivery among social vehicles:** As shown in Fig. 3.6, the content delivery among social vehicles in our framework mainly has two different cases: (1) vehicle i and vehicle ii have social ties; (2) vehicle i and vehicle ii do not have social ties. In the former case, when vehicle i encounters with vehicle ii, they will exchange content q downloaded from ECD j and content q' downloaded from ECD jj directly, where the above two contents have the strongest correlation. In the latter case, vehicle i and vehicle ii are strangers to each other. After vehicle i downloads the content from ECD j and encounters with vehicle ii, vehicle i may receive the request from vehicle ii to obtain content q. Then, vehicle i will send the content to vehicle ii, where the social ties between the above two vehicles are formed. By doing this, the latter case then becomes the former case, where vehicle ii may deliver content q' to vehicle i after the social ties are constructed. As such, instead of being a fixed customer to download content q' from ECD jj, vehicle i can obtain the content from social vehicles, where the traffic load of the ECD can be decreased.

3.5   Simulation

In this section, we evaluate the average transmission latency when a vehicle intends to download its requested content by comparing the proposed contract based framework with the conventional caching schemes. We first introduce the network setting followed by the results analysis.

3.5.1   Setting

We evaluate the average latency of content transmission in the coverage of an ECD by comparing our proposal with the random caching scheme (RCS) and the popularity based caching scheme (PBCS). In the RCS, the contents are cached and replaced in the ECD in a random way. As for the PBCS, the ECD selectively caches and replaces contents based on the popularity of contents. In the simulation, we divide the communication coverage of the ECD into 7 zones according to the fading channel, where the transmission rates of zones are $[1, 2, 5.5, 11, 5.5, 2, 1]$ Mbps [48]. The probability that a new coming vehicle is the fixed customer of the ECD is set to be 0.3. The content size is uniformly distributed with a range of $[0.1, 0.5]$ Mbytes. The caching capacity of the ECD is set to be 2.5 Mbytes. The number of contents in the networks is 100.

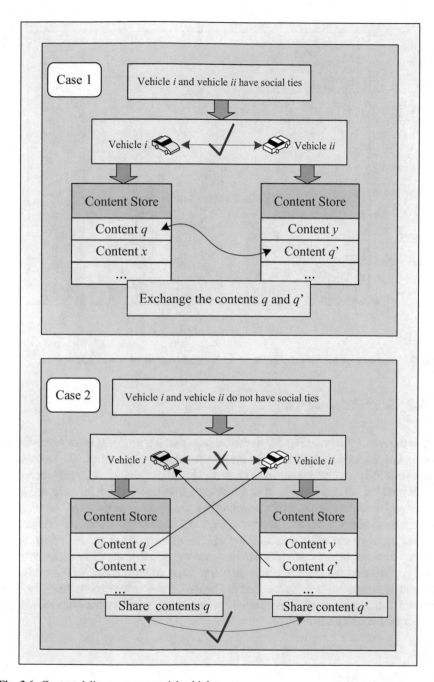

Fig. 3.6 Content delivery among social vehicles

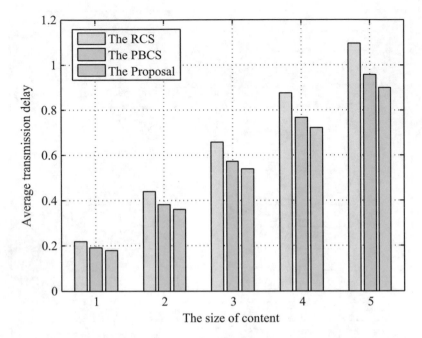

Fig. 3.7 The average transmission latency with different sizes of the requested content

3.5.2   Results Analysis

From the result shown in Fig. 3.7, it can be seen that the proposal leads to the lowest transmission latency compared with the two conventional schemes. The reasons for this are as follows. For one thing, the contract based caching strategy pre-caches the content which is signed by the fixed customers, leading to a high hit ratio and a low latency. For another thing, in the second part, the caching strategy of the proposal jointly considers the content popularity and content size to cache the contents to further improve the utilization of the caching capacity. As such, the transmission latency also can be decreased. In addition, we can see that the differences between the proposal and the other schemes become large when the size of the requested content is increased. It indicates that the proposal can reduce more time to obtain contents when the size of the requested content is larger, where the efficiency of the networks can be improved.

3.6   Summary

In this chapter, we have presented a novel contract based collaborative framework for edge caching in vehicular networks. By analyzing the traffic status and the requirements of vehicles, the contracts can be signed between ECDs and social vehicles in advance. Then, we have detailed the process of the contract based edge caching in vehicular networks including content caching, content replacement and content delivery. For each ECD, content caching and content replacement are based on three factors which are the contracts signed between vehicles and ECDs, the popularity of contents and the relevance among contents. As for the content delivery among social vehicles, we have considered two different cases according to their social ties. The proposed framework can decrease the transmission latency, enhance the efficiency of networks, and improve the QoE of social vehicles. The simulation result has verified the efficiency of the proposed framework with comparisons to the conventional schemes.

References

1. B. Ying, A. Nayak, A distributed social-aware location protection method in untrusted vehicular social networks. IEEE Trans. Veh. Technol. **68**(6), 6114–6124 (2019)
2. A.M. Vegni, V. Loscrì, A survey on vehicular social networks. IEEE Commun. Surv. Tutor. **17**(4), 2397–2419 (2015)
3. Z. Su, Y. Hui, S. Guo, D2D-based content delivery with parked vehicles in vehicular social networks. IEEE Wirel. Commun. **23**(4), 90–95 (2016)
4. X. Kong, F. Xia, Z. Ning, A. Rahim, Y. Cai, Z. Gao, J. Ma, Mobility dataset generation for vehicular social networks based on floating car data. IEEE Trans. Veh. Technol. **67**(5), 3874–3886 (2018)
5. C.A. Kerrache, N. Lagraa, R. Hussain, S.H. Ahmed, A. Benslimane, C.T. Calafate, J. Cano, A.M. Vegni, TACASHI: trust-aware communication architecture for social internet of vehicles. IEEE Internet Things J. **6**(4), 5870–5877 (2019)
6. B. Yu, F. Bai, PYRAMID: probabilistic content reconciliation and prioritization for V2V communications. IEEE Trans. Veh. Technol. **67**(7), 6615–6626 (2018)
7. Z. Zhou, H. Yu, C. Xu, Y. Zhang, S. Mumtaz, J. Rodriguez, Dependable content distribution in D2D-based cooperative vehicular networks: a big data-integrated coalition game approach. IEEE Trans. Intell. Transport. Syst. **19**(3), 953–964 (2018)
8. Y. Hui, Z. Su, T.H. Luan, J. Cai, Content in motion: an edge computing based relay scheme for content dissemination in urban vehicular networks. IEEE Trans. Intell. Transport. Syst. **20**(8), 3115–3128 (2019)
9. H. Li, B. Wang, Y. Song, K. Ramamritham, VeShare: a D2D infrastructure for real-time social-enabled vehicle networks. IEEE Wirel. Commun. **23**(4), 96–102 (2016)
10. C. Wu, T. Yoshinaga, X. Chen, L. Zhang, Y. Ji, Cluster-based content distribution integrating LTE and IEEE 802.11p with fuzzy logic and Q-learning. IEEE Comput. Intell. Mag. **13**(1), 41–50 (2018)
11. L. Gao, T.H. Luan, S. Yu, W. Zhou, B. Liu, FogRoute: DTN-based data dissemination model in fog computing. IEEE Internet Things J. **4**(1), 225–235 (2017)
12. L.T. Tan, R.Q. Hu, L. Hanzo, Twin-timescale artificial intelligence aided mobility-aware edge caching and computing in vehicular networks. IEEE Trans. Veh. Technol. **68**(4), 3086–3099 (2019)
13. Z. Su, Y. Hui, Q. Xu, T. Yang, J. Liu, Y. Jia, An edge caching scheme to distribute content in vehicular networks. IEEE Trans. Veh. Technol. **67**(6), 5346–5356 (2018)
14. Z. Zhou, H. Yu, C. Xu, Z. Chang, S. Mumtaz, J. Rodriguez, Begin: big data enabled energy-efficient vehicular edge computing. IEEE Commun. Mag. **56**(12), 82–89 (2018)

15. Y. Hui, Z. Su, T.H. Luan, J. Cai, Content in motion: an edge computing based relay scheme for content dissemination in urban vehicular networks. IEEE Trans. Intell. Transport. Syst. **20**(8), 3115–3128 (2019)

16. Y. Dai, D. Xu, S. Maharjan, G. Qiao, Y. Zhang, Artificial intelligence empowered edge computing and caching for internet of vehicles. IEEE Wirel. Commun. **26**(3), 12–18 (2019)

17. Z. Zhao, L. Guardalben, M. Karimzadeh, J. Silva, T. Braun, S. Sargento, Mobility prediction-assisted over-the-top edge prefetching for hierarchical vanets. IEEE J. Sel. Areas Commun. **36**(8), 1786–1801 (2018)

18. H. Zhou, N. Cheng, J. Wang, J. Chen, Q. Yu, X. Shen, Toward dynamic link utilization for efficient vehicular edge content distribution. IEEE Trans. Veh. Technol. **68**(9), 8301–8313 (2019)

19. Z. Su, Q. Xu, Q. Qi, Big data in mobile social networks: a QoE-oriented framework. IEEE Netw. **30**(1), 52–57 (2016)

20. Y. Meng, C. Jiang, T.Q.S. Quek, Z. Han, Y. Ren, Social learning based inference for crowdsensing in mobile social networks. IEEE Trans. Mobile Comput. **17**(8), 1966–1979 (2018)

21. E.K. Wang, Y. Li, Y. Ye, S.M. Yiu, L.C.K. Hui, A dynamic trust framework for opportunistic mobile social networks. IEEE Trans. Netw. Serv. Manag. **15**(1), 319–329 (2018)

22. X. Liu, D. He, C. Liu, Information diffusion nonlinear dynamics modeling and evolution analysis in online social network based on emergency events. IEEE Trans. Comput. Soc. Syst. **6**(1), 8–19 (2019)

23. J. Hu, L. Yang, L. Hanzo, Energy-efficient cross-layer design of wireless mesh networks for content sharing in online social networks. IEEE Trans. Veh. Technol. **66**(9), 8495–8509 (2017)

24. H. Ko, J. Lee, S. Pack, An opportunistic push scheme for online social networking services in heterogeneous wireless networks. IEEE Trans. Netw. Serv. Manag. **14**(2), 416–428 (2017)

25. T.H. Luan, R. Lu, X. Shen, F. Bai, Social on the road: enabling secure and efficient social networking on highways. IEEE Wirel. Commun. **22**(1), 44–51 (2015)

26. B. Ying, A. Nayak, A distributed social-aware location protection method in untrusted vehicular social networks. IEEE Trans. Veh. Technol. **68**(6), 6114–6124 (2019)

27. T.H. Luan, X. Shen, F. Bai, L. Sun, Feel bored? join verse! engineering vehicular proximity social networks. IEEE Trans. Veh. Technol. **64**(3), 1120–1131 (2015)

28. R. Yu, J. Kang, X. Huang, S. Xie, Y. Zhang, S. Gjessing, Mixgroup: accumulative pseudonym exchanging for location privacy enhancement in vehicular social networks. IEEE Trans. Depend. Secure Comput. **13**(1), 93–105 (2016)

29. T. Cheng, G. Liu, Q. Yang, J. Sun, Trust assessment in vehicular social network based on three-valued subjective logic. IEEE Trans. Multimedia **21**(3), 652–663 (2019)

30. X. Wang, Z. Ning, M. Zhou, X. Hu, L. Wang, Y. Zhang, F.R. Yu, B. Hu, Privacy-preserving content dissemination for vehicular social networks: challenges and solutions. IEEE Commun. Surv. Tutor. **21**(2), 1314–1345 (2019)

31. Z. Yang, H. Yu, J. Tang, H. Liu, Toward keyword extraction in constrained information retrieval in vehicle social network. IEEE Trans. Veh. Technol. **68**(5), 4285–4294 (2019)

32. R. Kim, H. Lim, B. Krishnamachari, Prefetching-based data dissemination in vehicular cloud systems. IEEE Trans. Veh. Technol. **65**(1), 292–306 (2016)

33. J. Wang, B. He, J. Wang, T. Li, Intelligent VNFs selection based on traffic identification in vehicular cloud networks. IEEE Trans. Veh. Technol. **68**(5), 4140–4147 (2019)

34. C. Lin, D. Deng, C. Yao, Resource allocation in vehicular cloud computing systems with heterogeneous vehicles and roadside units. IEEE Internet Things J. **5**(5), 3692–3700 (2018)

35. H.A. Khattak, H. Farman, B. Jan, I.U. Din, Toward integrating vehicular clouds with IoT for smart city services. IEEE Netw. **33**(2), 65–71 (2019)

36. K. Fan, X. Wang, K. Suto, H. Li, Y. Yang, Secure and efficient privacy-preserving ciphertext retrieval in connected vehicular cloud computing. IEEE Netw. **32**(3), 52–57 (2018)

37. M. Xing, J. He, L. Cai, Maximum-utility scheduling for multimedia transmission in drive-thru internet. IEEE Trans. Veh. Technol. **65**(4), 2649–2658 (2016)

38. W.L. Tan, W.C. Lau, O. Yue, T.H. Hui, Analytical models and performance evaluation of drive-thru internet systems. IEEE J. Sel. Areas. Commun. **29**(1), 207–222 (2011)
39. Z. Su, Y. Hui, T.H. Luan, S. Guo, Engineering a game theoretic access for urban vehicular networks. IEEE Trans. Veh. Technol. **66**(6), 4602–4615 (2017)
40. H. Zhou, B. Liu, F. Hou, T.H. Luan, N. Zhang, L. Gui, Q. Yu, X.S. Shen, Spatial coordinated medium sharing: optimal access control management in drive-thru internet. IEEE Trans. Intell. Transport. Syst. **16**(5), 2673–2686 (2015)
41. Y. Hui, Z. Su, T.H. Luan, J. Cai, A game theoretic scheme for optimal access control in heterogeneous vehicular networks. IEEE Trans. Intell. Transport. Syst. **20**(12), 4590–4603 (2019)
42. Z. Su, Y. Hui, Q. Yang, The next generation vehicular networks: a content-centric framework. IEEE Wirel. Commun. **24**(1), 60–66 (2017)
43. D. Wang, H. Cheng, P. Wang, X. Huang, G. Jian, Zipf's law in passwords. IEEE Trans. Inf. Forensics Secur. **12**(11), 2776–2791 (2017)
44. N. Golrezaei, A.G. Dimakis, A.F. Molisch, Scaling behavior for device-to-device communications with distributed caching. IEEE Trans. Inf. Theory **60**(7), 4286–4298 (2014)
45. A.S. Daghal, H. Zhu, J. Wang, Content delivery analysis in multiple devices to single device communications. IEEE Trans. Veh. Technol. **67**(11), 10218–10231 (2018)
46. H.A. Mustafa, M.Z. Shakir, M.A. Imran, R. Tafazolli, Spatial and social paradigms for interference and coverage analysis in underlay d2d network. IEEE Trans. Veh. Technol. **66**(10), 9328–9337 (2017)
47. M. Hajimirsadeghi, N.B. Mandayam, A. Reznik, Joint caching and pricing strategies for popular content in information centric networks. IEEE J. Sel. Areas Commun. **35**(3), 654–667 (2017)
48. T.H. Luan, X. Ling, X. Shen, Mac in motion: Impact of mobility on the mac of drive-thru internet. IEEE Trans. Mobile Comput. **11**(2), 305–319 (2012)

Chapter 4
Stackelberg Game Based Computation Offloading in Vehicular Networks

In the vehicular networks, the application of mobile edge computing (MEC) technology can assist road mobile vehicles to efficiently perform task computation and offloading services. From the perspective of the vehicles, it can save the time and energy consumption of computing tasks. From the perspective of the MEC servers, they can obtain benefits by providing task offloading services for vehicles. However, due to the limited computing power of the MEC servers and the need to consume network resources such as power and bandwidth for performing task computation, each MEC server typically cannot provide offloading services for all the vehicles within its coverage. In addition, for each MEC server in the networks, it needs to study an effective incentive mechanism in order to attract more vehicles to perform task offloading to obtain benefits. In order to solve the above issues, this chapter proposes a computation offloading scheme based on the Stackelberg game model. In the scheme, we first establish the Stackelberg game model of mobile vehicles and MEC server, where the benefit functions of them are designed. Then, by demonstrating that the vehicle and MEC server have the unique Stackelberg equilibrium solution, a distributed computation offloading algorithm based on Stackelberg game is designed to obtain the optimal game strategies for mobile vehicle and MEC server, respectively. Finally, the simulation results demonstrate the effectiveness of the proposed Stackelberg game based computation offloading scheme.

4.1 Introduction

The vehicular networks [1–5] are enriched with emerging applications such as vehicle detection, image processing, video streaming, speech recognition, augmented reality, etc. The applications generate enormous data traffic which is a challenge for the vehicular networks. Mobile edge computing (MEC) [6–10] enables a wide

© Springer Nature Switzerland AG 2021

Z. Su et al., *The Next Generation Vehicular Networks, Modeling, Algorithm, and Applications*, Wireless Networks, https://doi.org/10.1007/978-3-030-56827-6_4

range of vehicular application services to be offloaded and migrated to nearby edge servers so that the process of the application can be accelerated and the performance of the vehicular networks can be improved [11–15]. However, due to the dynamic topology of the network and the high-speed mobility of the vehicles, there are still many challenges for MEC-based computation offloading in vehicular networks.

- When the vehicular user decides to offload its computing tasks, both the user and the MEC server want to maximize their benefits. Therefore, it is extremely important to develop a reasonable offloading strategy in the vehicular networks.
- Based on the computation offloading problem, an incentive mechanism needs to be designed to promote the offloading transaction between the vehicular user and the MEC server. Therefore, the problem of MEC-based computation offloading in the vehicular networks requires an effective solution to facilitate the computing services requested by vehicles.

In the view of above problems, this chapter deeply studies the computation offloading technology based on the MEC architecture in the vehicular networks. By considering the characteristics of vehicles and the network, we design the optimal edge computing offloading scheme to improve the quality of experience (QoE) of vehicular users and the efficiency of edge offloading. The specific research contents are as follows:

- Considering the computation offloading problem in the vehicular networks, we design a network system based on MEC offloading and introduce the benefits of the vehicles and MEC servers. The proposal of the distributed computation offloading scheme makes that the vehicles can achieve the highest QoE and the MEC servers can obtain more benefits.
- In the MEC-based vehicular networks, the MEC server intends to attract more vehicles to offload their computation tasks. We thus study an incentive mechanism by designing the Stackelberg game model and apply the reverse induction method to analyze the computation offloading problem. With this mechanism, both the vehicle and the MEC server can be motivated to obtain the optimal game strategy.

4.2   System Model

4.2.1   Network Model

With the continuous increase of vehicles on the road and the enrichment of vehicular applications, the traditional backhaul networks cannot satisfy the various computing tasks of vehicles in real time (such as path planning, location sharing, augmented reality and other entertainment services) [16–20]. Compared with the cloud computing, MEC technology provides a new opportunity for vehicular computation offloading with its closer distance, lower latency, and more convenient computing

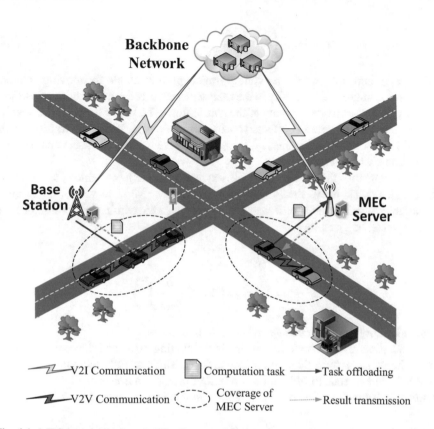

Fig. 4.1 MEC-based vehicle task offloading scenario

services [21–25]. Figure 4.1 shows a schematic diagram of the computation offloading scenario based on MEC. Considering the vehicle to vehicle (V2V) communication [26–28] and vehicle to infrastructure (V2I) communication [29–31], the MEC server is generally deployed at the roadside unit (RSU) and the base station (BS), providing edge computation offloading services for vehicles on the road, where the set of MEC servers is denoted by $\mathbb{M} = \{1, \ldots, m, \ldots, M\}$.

When a vehicle intends to communicate with the MEC server, the vehicle should enter the communication coverage of the MEC server. This chapter considers that each MEC server has the same coverage radius, denoted by R. According to [32], the average number of arriving vehicles moving within the communication coverage of MEC server m can be expressed as

$$N_m = \rho_m v_m, \tag{4.1}$$

where ρ_m and v_m respectively represent the traffic density and driving speed of vehicles within the coverage of the MEC server. Based on the studies in [33–35], the relationship between ρ_m and v_m can be expressed as

$$v_m = v_{\max} \left(1 - \left(\frac{\rho_m}{\rho_{\max}} \right)^\alpha \right), \alpha = 1, 2, 3, \ldots \tag{4.2}$$

where $v_{\max}$ and $\rho_{\max}$ are the maximum travel speed available for moving vehicles and the maximum traffic density in the road segment. α is the traffic flow adjustment factor that characterizes the state of the traffic flow. The larger the α is, the smaller the vehicle's travel speed is affected by the traffic density. When the traffic density reaches its maximum value, i.e., $\rho_m = \rho_{\max}$, we have $v_m = 0$. It means that the road is completely blocked.

Assuming that the computing task request generated by vehicle $i(i \in I = \{1, \ldots, i, \ldots, I\})$ obeys the Poisson distribution. For MEC server m, the number of vehicles which intend to offload the computing task within the communication coverage can be given by

$$N_m^C = \lambda_i \rho v$$

$$= \lambda_i \rho v_{\max} \left(1 - \left(\frac{\rho}{\rho_{\max}} \right)^\alpha \right), \tag{4.3}$$

where λ_i represents the average arrival rate of vehicles.

In addition, each vehicle has a certain computing power so that some vehicles may choose to perform their computing tasks locally. Let p_i^C denote the possibility that vehicle i intends to offload its computing task, the number of vehicles that are willing to offload their tasks to MEC server m then can be given by

$$N_m^{CO} = p_i^C \cdot \lambda_i \rho v$$

$$= p_i^C \lambda_i \rho v_{\max} \left(1 - \left(\frac{\rho}{\rho_{\max}} \right)^\alpha \right). \tag{4.4}$$

In this chapter, the coverage radius of each MEC server is independent. It means that each mobile vehicle belongs to only one MEC server, we have $N_m^{CO} \cap N_{m'}^{CO} = \emptyset, \forall m, m' \in M$.

4.2.2　Communication Model

When vehicle i enters the communication range of MEC server m, the communication connection is made through the RSU or BS in the vehicular networks. In general, the communication transmission between them is affected by the communication connection of other vehicles and external noise. Considering these factors comprehensively, according to Shannon's theorem, the data rate when vehicle i offloads its task to MEC server m can be calculated by

$$R_{i,m} = W_m \log_2 \left(1 + \frac{P_i H_{i,m}}{\sum_{j \in N_m^{CO}, j \neq i} P_j H_{j,m} + \sigma^2} \right), \tag{4.5}$$

where W_m represents the spectrum bandwidth allocated by the backhaul network to MEC server m. P_i is the transmission power density of vehicle i. $H_{i,m}$ is the channel gain between vehicle i and MEC server m. σ^2 represents the power spectral density of the background noise.

4.2.3 Task Execution Model

In vehicular networks, vehicle i has two choices to execute its computing task, namely local computing and computation offloading. Let $x_{i,m}$ denote the decision value, where $x_{i,m} = 1$ indicates that the computation task of vehicle i is offloaded to MEC server m; otherwise, $x_{i,m} = 0$. For vehicle i, its computing task can be expressed as $L_i \triangleq \left(Z_{i,k}^{in}, Z_{i,k}^{out}, D_i^k, t_{i,k}^{max} \right), i \in N_m$ [36–38]. Note that different vehicles generate different computing task requests and there are K types of computing tasks. $t_{i,k}^{max}$ is the maximum delay allowed for task k. $Z_{i,k}^{in}$ and $Z_{i,k}^{out}$ are the input size and output size of the computing task, respectively. D_i^k is the computing resources (i.e., the number of CPU cycles) required to complete the computing task. It has a direct relationship with the input size of the computing task, we have

$$D_i^k = h_k Z_{i,k}^{in}, \tag{4.6}$$

where h_k is a parameter related to the type of the computing task.

In what follows, we detail the analysis of the task execution model based on different task execution decisions.

(1) Local Computing Model
As a smart mobile device, each vehicle can execute the task by using the resources owned by itself. Compared with the computing power of MEC server, the computing resources owned by vehicles are limited. When vehicle i decides to locally execute its generated computing task, the average response time of the computing task can be expressed as

$$T_{i,k}^V = \frac{D_i^k}{f_{i,k}^V \left(1 - l_i^V \right)}, \tag{4.7}$$

where $f_{i,k}^V$ represents the computing power of vehicle i, that is, the CPU cycles that can be executed per unit time. l_i^V is the percentage of locally occupied computing resources.

According to the relationship between the time and the energy consumption, the energy consumption required for local task execution can be expressed as

$$E_{i,k}^{V} = \left(\alpha + \beta f_{i,k}^{V}{}^{3}\right) \frac{D_{i}^{k}}{f_{i,k}^{V}\left(1 - l_{i}^{V}\right)},$$ (4.8)

where α and β respectively represent the static power coefficient and dynamic power coefficient.

(2) Computation Offloading Model

When vehicle i chooses to offload its computing task to the MEC server, the task execution process mainly includes the transmission of the task and the computation of the task [39–43]. Specifically, vehicle i first transmits the generated computing task request through the wireless connection to its connected MEC server m. When MEC server m receives the tasks of vehicles within its coverage, computing resources will be allocated to each of these tasks. After these tasks are completed, MEC server m will return the computing results to the corresponding vehicles. During the execution of the offloaded tasks, the total transmission time of task L_i includes the input transmission time and the output transmission time of the task. It can be given by

$$
\begin{aligned}
T_{i,k,trans}^{MEC} &= \frac{Z_{i,k}^{in}}{R_{i,m}} + \frac{Z_{i,k}^{out}}{R_{i,m}} \\
&= \frac{Z_{i,k}^{in} + Z_{i,k}^{out}}{W_m \log_2 \left(1 + \frac{P_i H_{i,m}}{\sum_{j \in N_m^{CO}, j \neq i} P_j H_{j,m} + \sigma^2}\right)}.
\end{aligned}
$$ (4.9)

Next, we analyze the execution time of the offloaded task. Define the maximum task request acceptance rate of MEC server m as $\lambda_{\max}^{MEC}$. Within the communication range of MEC server m, the task computation request rate λ_{Sum}^{MEC} generated by all the vehicles can be expressed as

$$\lambda_{Sum}^{MEC} = \sum_{i=1}^{N_m} \lambda_i p_i^C.$$ (4.10)

Therefore, the execution ratio Γ^{MEC} of MEC server m to all the computing tasks can be expressed as

$$\Gamma^{MEC} = \begin{cases} 1, & \lambda_{Sum}^{MEC} \leq \lambda_{\max}^{MEC} \\ \frac{\lambda_{\max}^{MEC}}{\lambda_{Sum}^{MEC}}, & \lambda_{Sum}^{MEC} > \lambda_{\max}^{MEC} \end{cases}.$$ (4.11)

The execution rate of the computing tasks received by MEC server m is

$$
\begin{aligned}
\lambda_{exe}^{MEC} &= I_{\{x_{i,m}=1\}} \cdot \Gamma^{MEC} \lambda_{Sum}^{MEC} \\
&= \begin{cases}
I_{\{x_{i,m}=1\}} \cdot \lambda_{Sum}^{MEC}, & \lambda_{Sum}^{MEC} \leq \lambda_{max}^{MEC} \\
I_{\{x_{i,m}=1\}} \cdot \lambda_{max}^{MEC}, & \lambda_{Sum}^{MEC} > \lambda_{max}^{MEC}
\end{cases},
\end{aligned}
\tag{4.12}
$$

where $I_{\{\#\}}$ is an indicator function representing the task offloading decision, $I_{\{\#\}} = 1$ indicates that the offloading decision is true; otherwise, the decision is false. Then the execution time of task L_i is

$$
T_{i,k,exe}^{MEC} = \frac{D_i^k}{\lambda_{exe}^{MEC} f_{i,k}^{MEC}},
\tag{4.13}
$$

where $f_{i,k}^{MEC}$ is the computing power of MEC server m.

According to (4.9), (4.12) and (4.13), the average response time of the MEC server is

$$
\begin{aligned}
T_{i,k}^{MEC} &= T_{i,k,trans}^{MEC} + T_{i,k,exe}^{MEC} \\
&= \begin{cases}
\dfrac{Z_{i,k}^{in}+Z_{i,k}^{out}}{W_m \log_2\left(1+\dfrac{P_i H_{i,m}}{\sum_{j \in N_m^{CO}, j \neq i} P_j H_{j,m}+\sigma^2}\right)} + I_{\{x_{i,m}=1\}} \cdot \dfrac{D_i^k}{\lambda_{Sum}^{MEC} f_{i,k}^{MEC}}, & \lambda_{Sum}^{MEC} \leq \lambda_{max}^{MEC} \\[4mm]
\dfrac{Z_{i,k}^{in}+Z_{i,k}^{out}}{W_m \log_2\left(1+\dfrac{P_i H_{i,m}}{\sum_{j \in N_m^{CO}, j \neq i} P_j H_{j,m}+\sigma^2}\right)} + I_{\{x_{i,m}=1\}} \cdot \dfrac{D_i^k}{\lambda_{max}^{MEC} f_{i,k}^{MEC}}, & \lambda_{Sum}^{MEC} > \lambda_{max}^{MEC}
\end{cases}.
\end{aligned}
\tag{4.14}
$$

Let $\hat{\lambda}_{MEC} = \min\left(\lambda_{Sum}^{MEC}, \lambda_{max}^{MEC}\right)$, then (4.14) can be rewritten as

$$
\begin{aligned}
T_{i,k}^{MEC} &= \frac{Z_{i,k}^{in}+Z_{i,k}^{out}}{W_m \log_2\left(1+\dfrac{P_i H_{i,m}}{\sum_{j \in N_m^{CO}, j \neq i} P_j H_{j,m}+\sigma^2}\right)} \\
&\quad + I_{\{x_{i,m}=1\}} \cdot \frac{D_i^k}{\hat{\lambda}_{MEC} f_{i,k}^{MEC}}.
\end{aligned}
\tag{4.15}
$$

Correspondingly, the energy consumption during the task execution can be expressed as

$$
\begin{aligned}
E_{i,k}^{MEC} &= P_i T_{i,k}^{MEC} \\
&= P_i \cdot \frac{Z_{i,k}^{in}+Z_{i,k}^{out}}{W_m \log_2\left(1+\dfrac{P_i H_{i,m}}{\sum_{j \in N_m^{CO}, j \neq i} P_j H_{j,m}+\sigma^2}\right)} \\
&\quad + I_{\{x_{i,m}=1\}} \cdot \frac{P_i D_i^k}{\hat{\lambda}_{MEC} f_{i,k}^{MEC}}.
\end{aligned}
\tag{4.16}
$$

4.3   Stackelberg Game Analysis

This section uses the Stackelberg game [44–48] to analyze the interaction of the task offloading between a vehicle and a MEC server. We first define the benefit functions of the vehicle and the MEC server, respectively. Then, we analyze the Stackelberg game and use the inverse induction method to find the optimal strategy of both parties in the game.

4.3.1   Benefits of Vehicles

When vehicle i chooses to offload its computing task to MEC server m, both the task completion time and the resource energy consumption of the task can be reduced. Therefore, the benefits of vehicle i can be defined as the task completion time and the energy saved by the MEC server minus the payment for the task offloading service, shown as

$$U_{i,k}^V = F\left(T_{i,k}^V, T_{i,k}^{MEC}, E_{i,k}^V, E_{i,k}^{MEC}\right) - P_{MEC}(p_{MEC}), \qquad (4.17)$$

where $F\left(T_{i,k}^V, T_{i,k}^{MEC}, E_{i,k}^V, E_{i,k}^{MEC}\right)$ represents the satisfaction function of vehicle i, that is, the sum of time and energy saved by offloading the computing task to the MEC server. $P_{MEC}(p_{MEC})$ represents the cost of the vehicle, i.e., the price that needs to be paid for the MEC server to perform the service. p_{MEC} is the price unit computing resource.

The time and energy consumption saved by offloading the task can be given by

$$\Delta T_{i,k} = T_{i,k}^V - T_{i,k}^{MEC}. \qquad (4.18)$$

$$\Delta E_{i,k} = E_{i,k}^V - E_{i,k}^{MEC}. \qquad (4.19)$$

As such, the satisfaction function of vehicle i can be defined by

$$F(T_{i,k}^V, T_{i,k}^{MEC}, E_{i,k}^V, E_{i,k}^{MEC}) = \eta \Delta T_{i,k} + (1 - \eta)\Delta E_{i,k}, \qquad (4.20)$$

where $0 \leq \eta \leq 1$, which represents the preference factor of vehicle i's satisfaction with the offloading service. When vehicle i pays more attention to the time, the value of η is larger than 0.5; otherwise, we have $\eta < 0.5$.

For the payment, it is related to the price unit resource, shown as

$$P_{MEC}(p_{MEC}) = p_{MEC}\left(f_{i,k}^{MEC} - \zeta\right), i \in N_m, \qquad (4.21)$$

where p_{MEC} is the price unit resource when the MEC server provides the offloading service for computing task L_i requested by vehicle i.

By substituting (4.7), (4.8), (4.15), (4.16), (4.20) and (4.21) into (4.17), the benefits of vehicle i can be rewritten as

$$U_{i,k}^V = \eta \Delta T_{i,k} + (1 - \eta) \Delta E_{i,k} - p_{MEC} \cdot \left(f_{i,k}^{MEC} - \zeta \right) \tag{4.22}$$

$$= \eta \left(\frac{D_i^k}{f_{i,k}^V (1 - l_i^V)} - I_{\{x_{i,m}=1\}} \cdot \frac{D_i^k}{\hat{\lambda}_{MEC} f_{i,k}^{MEC}} \right.$$

$$\left. - \frac{Z_{i,k}^{in} + Z_{i,k}^{out}}{W_m \log_2 \left(1 + \frac{P_i H_{i,m}}{\sum_{j \in N_m^{CO}, j \neq i} P_j H_{j,m} + \sigma^2} \right)} \right) + (1-\eta) \left(\left(\alpha + \beta f_{i,k}^{V\,3} \right) \frac{D_i^k}{f_{i,k}^V (1 - l_i^V)} \right.$$

$$\left. - I_{\{x_{i,m}=1\}} \cdot \frac{P_i D_i^k}{\hat{\lambda}_{MEC} f_{i,k}^{MEC}} - P_i \frac{Z_{i,k}^{in} + Z_{i,k}^{out}}{W_m \log_2 \left(1 + \frac{P_i H_{i,m}}{\sum_{j \in N_m^{CO}, j \neq i} P_j H_{j,m} + \sigma^2} \right)} \right)$$

$$- p_{MEC} \left(f_{i,k}^{MEC} - \zeta \right).$$

In order to ensure that the benefits of vehicle i is a positive value, we have $U_{i,k}^V > 0$.

4.3.2 Benefits of MEC Server

In the task offloading process, the strategy of MEC server is to determine the optimal price unit computing resource to provide offloading services for vehicles to obtain more profits. When a vehicle decides to offload its computing task to the MEC server, the MEC server will obtain a certain payment for completing the task, which can be regarded as the income of the MEC server. At the same time, the MEC server needs to consume the computing resources to complete the task. The benefits of the MEC server then can be defined as

$$U_{MEC} = \hat{F}(p_{MEC}, f_{i,k}^{MEC}) - C_{MEC}(\mu_{MEC}), \tag{4.23}$$

where $\hat{F}(p_{MEC}, f_{i,k}^{MEC})$ is the benefit obtained by MEC server from vehicle i. $C_{MEC}(\mu_{MEC})$ represents the computing cost of the MEC server to complete the computing task. Since one MEC server can provide offloading services for multiple vehicles within its communication coverage, the MEC server's offloading revenue can be expressed as

$$\hat{F}(p_{MEC}, f_{i,k}^{MEC}) = \sum_{k=1}^{K} \sum_{i=1}^{N_m^{CO}} p_{MEC} \left(f_{i,k}^{MEC} - \zeta \right), \qquad (4.24)$$

where ζ represents the subsidy of the extra computing resources provided by the MEC server to motivate vehicles to offload their computing tasks. The more times a vehicle requests task computing service from the MEC server, the more subsidies the MEC server gives.

Considering the limited computing power of MEC server, the cost of the service can be expressed as

$$C_{MEC}(\mu_{MEC}) = \mu_{MEC} \cdot \min \left(\lambda_{Sum}^{MEC} \sum_{k=1}^{K} \sum_{i=1}^{N_m^{CO}} f_{i,k}^{MEC}, \lambda_{max}^{MEC} \sum_{k=1}^{K} \sum_{i=1}^{N_m^{CO}} f_{i,k}^{MEC} \right),$$

$$(4.25)$$

where μ_{MEC} is the cost of providing unit computing resource for the offloading service. $\lambda_{Sum}^{MEC} \sum_{i=1}^{N_m^{CO}} f_{i,k}^{MEC}$ represents the computing resources provided by MEC server for vehicles within its communication coverage, which should be no larger than its maximum computing resources $\lambda_{max}^{MEC} \sum_{i=1}^{N_m^{CO}} f_{i,k}^{MEC}$.

By substituting (4.24) and (4.25) into (4.23), the benefits of MEC server can be rewritten as

$$U_{MEC} = \sum_{k=1}^{K} \sum_{i=1}^{N_m^{CO}} p_{MEC} \left(f_{i,k}^{MEC} - \zeta \right) - \mu_{MEC} \cdot \hat{\lambda}_{MEC} \sum_{k=1}^{K} \sum_{i=1}^{N_m^{CO}} f_{i,k}^{MEC}.$$

$$(4.26)$$

4.3.3   Stackelberg Game

In this section, we model the pricing of MEC server and the decisions of vehicles as a non-cooperative Stackelberg game based on their benefit functions. After this, we analyze the optimal strategies of the two players to achieve the game equilibrium. When a vehicle plans to offload its computing task to the nearby MEC server, it will send the task information such as the input size of the task and the computing resources required to complete the task to the MEC server. The MEC server then calculates the optimal pricing strategy to maximize benefits based on its own computing power and the cost of executing the task and sends the final price to the vehicle within its communication coverage. Finally, the vehicle determines the best strategy for calculating the offloading service based on the price declared by the MEC server.

In the Stackelberg game, the MEC server is the leader in the game process. In the first stage of the game, the price unit resource μ_{MEC} is determined by the MEC server. Based on the price, the vehicle then determines the number of computing resources to request from the MEC server. Therefore, the vehicle is the follower in the Stackelberg game.

In general, both the vehicles and the MEC server are rational individuals in the game, and each of them intends to maximize its benefits. For vehicles, the problem can be expressed as

Problem 4.1

$$\arg \max_{f_{i,k}^{MEC}} U_{i,k}^{V}. \tag{4.27}$$

Similarly, for MEC server, the problem can be formulated as

Problem 4.2

$$\arg \max_{p_{MEC}} U_{MEC} = \sum_{k=1}^{K} \sum_{i=1}^{N_m^{CO}} p_{MEC} \left(f_{i,k}^{MEC} - \zeta \right) - \mu_{MEC} \cdot \hat{\lambda}_{MEC} \sum_{k=1}^{K} \sum_{i=1}^{N_m^{CO}} f_{i,k}^{MEC}.$$

$$s.t. \ p_{MEC} > \mu_{MEC}.$$

$$\tag{4.28}$$

4.4 The Equilibrium Solution of Stackelberg Game

In this subsection, the inverse inductive method is used to solve the equilibrium solution of the Stackelberg game. We first consider the vehicle and obtain its task offloading strategy in the second stage of the game. Then, the MEC server determines the optimal pricing of the offloading service to maximize the benefit in the first stage. Based on this, a distributed computation offloading algorithm is designed to obtain the equilibrium solution of the game.

4.4.1 Stage 2: Offloading Strategy of Vehicles

When the MEC server gives the price of the unit computing resource for performing the task, the vehicle needs to determine the number of computing resources to obtain from the MEC server. To this end, we find the first derivative of $U_{i,k}^{V}$ with respect to $f_{i,k}^{MEC}$, shown as

$$\frac{\partial U_{i,k}^V}{\partial f_{i,k}^{MEC}} = I_{\{x_{i,m}=1\}} \cdot \frac{\eta D_i^k + (1-\eta) P_i D_i^k}{\hat{\lambda}_{MEC}} \cdot \left(\frac{1}{f_{i,k}^{MEC}}\right)^2 - p_{MEC}. \tag{4.29}$$

Furthermore, the second derivative of $U_{i,k}^V$ can be expressed as

$$\frac{\partial^2 U_{i,k}^V}{\partial f_{i,k}^{MEC2}} = -I_{\{x_{i,m}=1\}} \cdot \frac{2\eta D_i^k + 2(1-\eta) P_i D_i^k}{\hat{\lambda}_{MEC}} \cdot \left(\frac{1}{f_{i,k}^{MEC}}\right)^3 . \tag{4.30}$$

In (4.30), all the parameters are larger than zero, so the second derivative of $U_{i,k}^V$ with respect to $f_{i,k}^{MEC}$ is less than 0, i.e., $\partial^2 U_{i,k}^V / \partial f_{i,k}^{MEC2} < 0$. It means that the first derivative of $U_{i,k}^V$ is strictly decreased with $f_{i,k}^{MEC}$. We have

$$\begin{cases} \lim_{f_{i,k}^{MEC} \to 0} \frac{\partial u_{i,k}^V}{\partial f_{i,k}^{MEC}} = +\infty > 0 \\ \lim_{f_{i,k}^{MEC} \to \infty} \frac{\partial u_{i,k}^V}{\partial f_{i,k}^{MEC}} = -p_{MEC} < 0 \end{cases}. \tag{4.31}$$

This formula shows that the benefit of the vehicle $U_{i,k}^V$ increases with the increase of the computing resources allocated by the MEC server, and therefore, the benefit function $U_{i,k}^V$ of the vehicle is a strict concave function with respect to $f_{i,k}^{MEC}$. According to Fermat's theorem, if $U_{i,k}^V / f_{i,k}^{MEC} = 0$, the vehicle can obtain the optimal offloading strategy, shown as

$$R_V\left(p_{MEC}^*\right) = f_{i,k}^{MFC*} = I_{\{x_{i,m}=1\}} \sqrt{\frac{\eta D_i^k + (1-\eta) P_i D_i^k}{\hat{\lambda}_{MEC} p^{MEC}}}, \tag{4.32}$$

where $R_V\left(p_{MEC}^*\right)$ is the realistic response of the vehicle to the price of computing resources determined by the MEC server.

4.4.2 Stage 1: Pricing Strategy of MEC Servers

In the first stage of the Stackelberg game, the strategy of the MEC server is to calculate the price when the offloading resources are determined by vehicles. When giving the vehicle's offloading strategy, (4.28) can be rewritten as

$$\max_{p_{MEC}} U_{MEC}\left(p_{MEC}, R_V\left(p_{MEC}\right)\right)$$

$$= \sum_{k=1}^{K} \sum_{i=1}^{N_m^{CO}} I_{\{x_{i,m}=1\}} \cdot \left(p_{MEC} - \hat{\lambda}_{MEC} \cdot \mu_{MEC}\right) \sqrt{\frac{\eta D_i^k + (1-\eta) P_i D_i^k}{\hat{\lambda}_{MEC} p_{MEC}}} - p_{MEC}\varsigma. \tag{4.33}$$

In order to obtain the extreme value of the benefit function, we find the first derivative of U_{MEC} with respect to p_{MEC}, shown as

$$
\frac{\partial U_{MEC}}{\partial p_{MEC}} = \sum_{k=1}^{K} \sum_{i=1}^{N_m^{CO}} I_{\{x_{i,m}=1\}} \left(\frac{1}{2} \sqrt{\frac{\eta D_i^k + (1-\eta) P_i D_i^k}{\hat{\lambda}_{MEC}}} p_{MEC}^{-1/2} \right.
$$

$$
\left. \left(1 + \hat{\lambda}_{MEC} \mu_{MEC} p_{MEC}^{-1} \right) - \zeta \right). \tag{4.34}
$$

Let $\hat{p}_{MEC} = p_{MEC}^{-1/2}$, $\varXi = \sqrt{\frac{\eta D_i^k + (1-\eta) P_i D_i^k}{\hat{\lambda}_{MEC}}}$, then (4.34) becomes

$$
\frac{\partial U_{MEC}}{\partial p_{MEC}} = \sum_{k=1}^{K} \sum_{i=1}^{N_m^{CO}} I_{\{x_{i,m}=1\}} \left(\frac{1}{2} \varXi \hat{\lambda}_{MEC} \mu_{MEC} \hat{p}_{MEC}^3 + \frac{1}{2} \varXi \hat{p}_{MEC} - \zeta \right). \tag{4.35}
$$

Let $\partial U_{MEC}/\partial p_{MEC} = 0$, we then can obtain the optimal pricing strategy $\hat{p}_{MEC}^*$ under the constraint condition of satisfying $p_{MEC} > \mu_{MEC} > 0$, i.e. $0 < \hat{p}_{MEC} < 1/\sqrt{\mu_{MEC}}$. Therefore, the optimal pricing strategy of the MEC server can be expressed as

$$
R_{MEC}\left(p_{MEC}^*\right) = \frac{1}{\left(\hat{p}_{MEC}^*\right)^2} = p_{MEC}^*. \tag{4.36}
$$

4.4.3   Stackelberg Game Equilibrium

After analyzing the optimal pricing strategy and offloading strategy of the MEC server (leader) and the vehicle (follower) in the above-mentioned Stackelberg game, we then find the equilibrium of the game and prove the uniqueness of the equilibrium solution.

Theorem 4.1 *The vehicles and the MEC server have equilibrium solutions in the Stackelberg game.*

Proof We first give a conclusion that if the following conditions are satisfied, there is an equilibrium solution in the game.

(1) The price unit computing resource p_{MEC} is a non-empty, closed, bounded convex set in the Euclidean space R^N;
(2) $U_{MEC}(p_{MEC})$ is continuous and concave on the p_{MEC}.

In the Stackelberg game, the computing resources $f_{i,k}^{MEC}$ required by the vehicle is a non-empty, convex and tight subset of the Euclidean space. According to (4.29) and (4.30), the benefit function of the vehicle $U_{i,k}^V$ is a concave function with respect to $f_{i,k}^{MEC}$. Therefore, vehicles have an equilibrium solution in the Stackelberg game.

Similarly, for the MEC server, the price p_{MEC} is a non-empty, convex and tight subset of Euclidean space, and $U_{MEC}(p_{MEC})$ is continuous on p_{MEC}. According to Eq. (4.35), the second derivative of the benefit function $U_{MEC}(p_{MEC})$ of the MEC server with respect to the price p_{MEC} can be given by

$$\frac{\partial^2 U_{MEC}}{\partial p_{MEC}} = \sum_{k=1}^{K} \sum_{i=1}^{N_m^{CO}} I_{\{x_{i,m}=1\}} \cdot \left(-\frac{1}{4}\Xi\right) \left(\hat{p}_{MEC}^3 + 3\hat{\lambda}_{MEC}\mu_{MEC}\hat{p}_{MEC}^2\right) < 0.$$

$$(4.37)$$

Therefore, $\partial U_{MEC}/\partial p_{MEC}$ is strictly decreased with p_{MEC}.
Let $p_{MEC} \rightarrow \mu_{MEC}$, we have

$$\lim_{p_{MEC} \rightarrow \mu_{MEC}} \frac{\partial U_{MEC}}{\partial p_{MEC}} = \sum_{k=1}^{K} \sum_{i=1}^{N_m^{CO}} I_{\{x_{i,m}=1\}} \cdot \left(\frac{\Xi\left(\hat{\lambda}_{MEC+1}\right)}{2\sqrt{\mu_{MEC}}} - \zeta\right). \qquad (4.38)$$

Here, if $\frac{\Xi\left(\hat{\lambda}_{MEC+1}\right)}{2\sqrt{\mu_{MEC}}} > \zeta$, then $\lim_{p_{MEC} \rightarrow \mu_{MEC}} \frac{\partial U_{MEC}}{\partial p_{MEC}} > 0$.
Similarly, let $\hat{p}_{MEC} \rightarrow 1/\sqrt{\mu_{MEC}}$, we have

$$\lim_{p_{MEC}+\infty} \frac{\partial u_{MEC}}{\partial p_{MEC}} = \sum_{k=1}^{K} \sum_{i=1}^{N_m^{CO}} I_{\{x_{i,m}=1\}} \cdot (-\zeta) < 0. \qquad (4.39)$$

Therefore, the benefit function of the MEC server is first strictly increased and then strictly decreased. It means that the benefit function is a concave function. That is to say, the MEC server has an equilibrium solution in the Stackelberg game. Theorem 4.1 has been proved.

Theorem 4.2 *The vehicle has a unique Stackelberg equilibrium solution in the game.*

Proof It can be seen from Theorem 4.1 that the vehicle has an equilibrium solution in the Stackelberg game. Let $R_V\left(p_{MEC}^*\right) = f_{i,k}^{MEC*}$ be the optimal offloading strategy for the vehicle which satisfies the optimal response function Eq. (4.32). If $R_V\left(p_{MEC}^*\right)$ is a standard function, the Stackelberg equilibrium solution is unique [49]. Therefore, we first prove Eq. (4.32) is the standard function when the following conditions are met:

(1) Positivity: $R_V\left(p_{MEC}^*\right) > 0$;
(2) Monotonicity: If $p_{MEC}^* \leq \tilde{p}_{MEC}$, then $R_V\left(p_{MEC}^*\right) \leq R_V\left(\tilde{p}_{MEC}^*\right)$ or $R_V\left(p_{MEC}^*\right) \geq R_V\left(\tilde{p}_{MEC}^*\right)$;
(3) Scalability: For any $\epsilon \geq 1$, $\epsilon R_V\left(p_{MEC}^*\right) \geq R_V\left(p_{MEC}^*\right)$.

As a rational game player, the vehicle has $f_{i,k}^{MEC*} \geq 0$. Thus, the best response function $R_V\left(p_{MEC}^*\right)$ satisfies the positivity property. For the best response function $R_V\left(p_{MEC}^*\right)$, the first derivative of p_{MEC}^* is

$$\frac{\partial R_V\left(p_{MEC}^*\right)}{\partial p_{MEC}^*} = -\frac{1}{2}I_{\{x_{i,m}=1\}} \cdot \varXi \cdot \frac{1}{p_{MEC}\sqrt{p_{MEC}}} < 0. \tag{4.40}$$

From Eq. (4.40), we can know that $R_V\left(p_{MEC}^*\right)$ is monotonically decreasing with p_{MEC}^*. That is, if $p_{MEC}^* \leq \tilde{p}_{MEC}$, we have $R_V\left(p_{MEC}^*\right) \geq R_V\left(\tilde{p}_{MEC}^*\right)$. Therefore, the monotonicity is proved.

Next, we prove the scalability. Define $\varepsilon \geq 1$, we have

$$\frac{R_V\left(\epsilon p_{MEC}^*\right)}{\varepsilon R_V\left(p_{MEC}\right)} = \frac{I_{\{x_i,m=1\}} \cdot \varXi \left(\epsilon p_{MEC}\right)^{-1/2}}{I_{\{x_i,m=1\}} : \varXi \epsilon p_{MEC}^{-1/2}} = \frac{1}{\epsilon\sqrt{\epsilon}} < 1. \tag{4.41}$$

Therefore, the scalability is proved.

Since $R_V\left(p_{MEC}^*\right)$ is a standard function, the vehicles have a unique Stackelberg equilibrium solution in the game. Theorem 4.2 is proved.

According to the above analysis, a distributed computation offloading algorithm based on the Stackelberg game is designed. The specific implementation process is shown in Algorithm 4.1. In the algorithm, ℓ is a small value, such as $\ell = 10^{-4}$. v and θ are the step size parameters when the MEC server updates the price.

Algorithm 4.1 Distributed offloading algorithm based on Stackelberg game

---

1: **Initialization** The number of vehicles N_m^{CO};
2: The computation tasks $\left\{L_i \triangleq \left(Z_{i,k}^{in}, Z_{i,k}^{out}, D_i^k, t_{i,k}^{max}\right), i \in N_m, k \in K\right\}$;
3: The execution rate of MEC server $\hat{\lambda}_{MEC}$;
4: For $t = 0$, MEC server announces the price $p_{MEC}(t)$;
5: Repeat for $t = 1, 2, 3, \cdots$;
6: **for** vehicle $i = 1, 2, \cdots, N_m^{CO}$ **do**
7:      Determine the optimal resource demand $f_{i,k}^{MEC*}(t)$ using Eq. (4.32);
8: **end for**
9: **for** MEC server $m = 1, 2, \cdots, M$ **do**
10:     **if** $\hat{\lambda}_{MEC} = \lambda_{max}^{MEC}$ **then**
11:         $p_{MEC}(t+1) = p_{MEC}(t) + v\nabla U_{MEC}(p_{MEC})$;
12:     **else**
13:         $p_{MEC}(t+1) = p_{MEC}(t) - \theta p_{MEC}(t)$;
14:     **end if**
15: **end for**
16: **until** $|p_{MEC}(t+1) - p_{MEC}(t)|/p_{MEC}(t) < \ell$;
17: **until** $f_{i,k}^{MEC}(t+1) = f_{i,k}^{MEC}(t)$;

4.5   Simulation

This section evaluates the efficiency of the proposed Stackelberg game-based distributed computation offloading scheme by simulation. We first introduce the simulation experiment setting, followed by the analysis of the simulation results.

4.5.1   Setting

In the simulation, the computing task of the vehicle is performed by the MEC server. Since the tasks of different vehicles are not the same, the number of computing resources required to complete different tasks are randomly selected from $[2, 4, 6, 8, 10]$. The computing power of the vehicles is 5. Namely, we have $f_{i,k}^V = 5$. The value of ζ is set to be 1.5 in the simulation. The initial price is $p_{MEC} = 0.05$.

4.5.2   Results Analysis

In the Stackelberg game, the optimal offloading strategy of vehicles is to determine the number of computing resources that request from the MEC server $f_{i,k}^{MEC}$, and the strategy of the MEC server is to determine the task execution price p_{MEC}. Figure 4.2 shows the optimal offloading strategy of the vehicle with different task execution prices charged by the MEC server. When the price unit computing resource of the MEC server is lower, the vehicle is more willing to perform the edge computation offloading service. As a result, more computing resources of the MEC server will be allocated.

By offloading the task to the MEC server, the vehicle can save time and energy for performing local computing, i.e., the vehicle can obtain certain offloading benefits, as shown in Fig. 4.3. When the MEC server performs the task with a lower price, the vehicle is more willing to offload its task to the MEC server, thereby obtaining more offloading benefits.

In order to verify the efficiency of the Stackelberg game-based distributed computation offloading scheme proposed in this chapter, it is compared with two traditional computation offloading schemes. (1) Computation of the fixed price: In this scheme, the price of the computing resource provided by the MEC server for the offloading service is fixed; (2) Computation of the random price: In this scheme, the price of the computing resources is randomly varied within the price range. Figure 4.4 shows the offloading benefit of the vehicle $U_{i,k}^V$ by changing the number of computing resources D_i^k requested by the task. When the value of D_i^k is small, the differences of the offloading benefits obtained by vehicles in the three different schemes are small. However, when the value of D_i^k is gradually increased, the

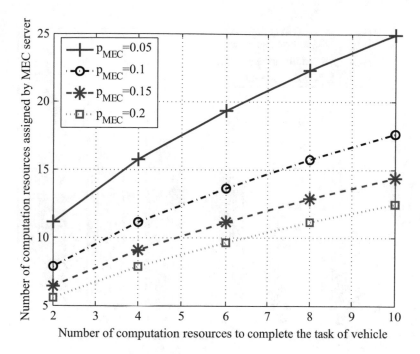

Fig. 4.2 Number of computing resources assigned by the MEC server with different prices and number of computation resources to complete the task

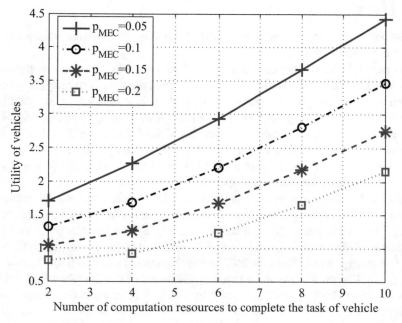

Fig. 4.3 The benefits of the vehicle with different prices and number of computation resources to complete the task

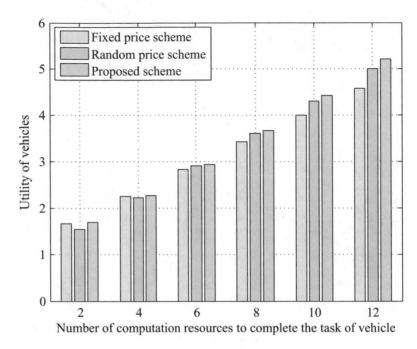

Fig. 4.4 The benefits of the vehicle by changing the number of computing resources requested by the task

computation offloading scheme proposed in this chapter can make vehicles obtain more offloading benefits.

Figure 4.5 shows the benefits of the MEC server by providing computation offloading services for vehicles in three different schemes. Obviously, the scheme proposed in this chapter is always the optimal. This is because in the fixed price scheme and the random price scheme, the price for executing the offloading task is too high or too low. As a result, the benefits of the MEC server is reduced. In the proposed scheme, the price of the offloading resources dynamically changes with the strategies of vehicles. Therefore, the MEC server can obtain the optimal benefits by providing task computing services.

Figure 4.6 shows the benefits obtained by the vehicle and the MEC server by changing the number of resources required for the completion of the computing task. When the value of D_i^k is small (e.g., $D_i^k \leq 8$), the benefits obtained by the vehicle are higher than that obtained by the MEC server. In contrast, when the value of D_i^k is large, the benefits of the MEC server become high. In addition, as shown in this figure, the growth rate of the benefits obtained by the vehicle is slightly slower than that of the MEC server. The reason for this is the task execution time increases with the increase of the computation resources required for the completion of the computing task. As a result, the QoE of vehicles will gradually increase, and the growth will be slower than that of the MEC server.

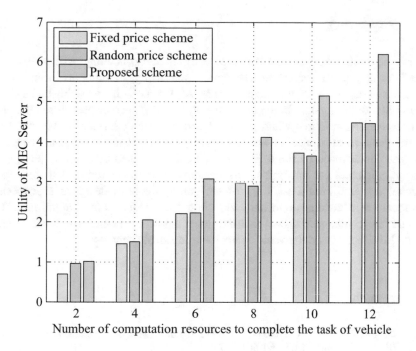

Fig. 4.5 The benefits of MEC server by providing computation offloading services for vehicles

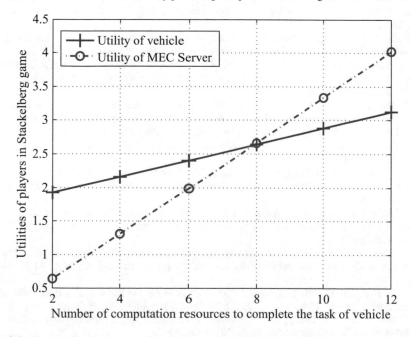

Fig. 4.6 The benefits obtained by the vehicle and the MEC server by changing the number of resources required for the completion of the computing task

4.6 Summary

This chapter has proposed a distributed computation offloading scheme based on Stackelberg game in vehicular networks. Specifically, in order to formulate the relationship between the vehicle and the MEC server, we have established a model of the Stackelberg game. In the game model, the MEC server determines the price unit computing resource when providing the offloading service in the first stage of the game to maximize its benefits. In the second stage of the game, based on the price determined by the MEC server, the vehicle then determines the optimal resource demand for maximizing its benefits. By demonstrating that there exist Stackelberg equilibrium solutions of the game, we have designed a distributed computation offloading algorithm to obtain the unique game equilibrium. The simulation results have shown that the proposed scheme can bring more benefits to the MEC server and the vehicle than the traditional schemes.

References

1. Z. Su, Y. Hui, Q. Yang, The next generation vehicular networks: a content-centric framework. IEEE Wirel. Commun. **24**(1), 60–66 (2017)
2. H. Peng, Q. Ye, X.S. Shen, SDN-based resource management for autonomous vehicular networks: a multi-access edge computing approach. IEEE Wirel. Commun. **26**(4), 156–162 (2019)
3. Z. Su, Y. Hui, Q. Xu, T. Yang, J. Liu, Y. Jia, An edge caching scheme to distribute content in vehicular networks. IEEE Trans. Veh. Technol. **67**(6), 5346–5356 (2018)
4. Y. Zhang, C. Li, T.H. Luan, Y. Fu, W. Shi, L. Zhu, A mobility-aware vehicular caching scheme in content centric networks: model and optimization. IEEE Trans. Veh. Technol. **68**(4), 3100–3112 (2019)
5. Y. Hui, Z. Su, T.H. Luan, J. Cai, A game theoretic scheme for optimal access control in heterogeneous vehicular networks. IEEE Trans. Intell. Transp. Syst. **20**(12), 4590–4603 (2019)
6. J. Zhao, Q. Li, Y. Gong, K. Zhang, Computation offloading and resource allocation for cloud assisted mobile edge computing in vehicular networks. IEEE Trans. Veh. Technol. **68**(8), 7944–7956 (2019)
7. K. Zhang, Y. Mao, S. Leng, Y. He, Y. Zhang, Mobile-edge computing for vehicular networks: a promising network paradigm with predictive off-loading. IEEE Veh. Technol. Mag. **12**(2), 36–44 (2017)
8. J. Liu, J. Wan, B. Zeng, Q. Wang, H. Song, M. Qiu, A scalable and quick-response software defined vehicular network assisted by mobile edge computing. IEEE Commun. Mag. **55**(7), 94–100 (2017)
9. Z. Ning, X. Wang, J. Huang, Mobile edge computing-enabled 5G vehicular networks: toward the integration of communication and computing. IEEE Veh. Technol. Mag. **14**(1), 54–61 (2019)
10. J. Feng, Z. Liu, C. Wu, Y. Ji, Mobile edge computing for the internet of vehicles: offloading framework and job scheduling. IEEE Veh. Technol. Mag. **14**(1), 28–36 (2019)
11. G. Qiao, S. Leng, K. Zhang, Y. He, Collaborative task offloading in vehicular edge multi-access networks. IEEE Commun. Mag. **56**(8), 48–54 (2018)
12. Z. Ding, J. Xu, O.A. Dobre, H.V. Poor, Joint power and time allocation for noma–mec offloading. IEEE Trans. Veh. Technol. **68**(6), 6207–6211 (2019)

13. B. Gu, Z. Zhou, Task offloading in vehicular mobile edge computing: a matching-theoretic framework. IEEE Veh. Technol. Mag. **14**(3), 100–106 (2019)
14. Y. Sun, L. Xu, Y. Tang, W. Zhuang, Traffic offloading for online video service in vehicular networks: a cooperative approach. IEEE Trans. Veh. Technol. **67**(8), 7630–7642 (2018)
15. Y. Wu, L.P. Qian, H. Mao, X. Yang, H. Zhou, X. Tan, D.H.K. Tsang, Secrecy-driven resource management for vehicular computation offloading networks. IEEE Netw. **32**(3), 84–91 (2018)
16. Z. Zhou, P. Liu, J. Feng, Y. Zhang, S. Mumtaz, J. Rodriguez, Computation resource allocation and task assignment optimization in vehicular fog computing: a contract-matching approach. IEEE Trans. Veh. Technol. **68**(4), 3113–3125 (2019)
17. Y. Sun, X. Guo, J. Song, S. Zhou, Z. Jiang, X. Liu, Z. Niu, Adaptive learning-based task offloading for vehicular edge computing systems. IEEE Trans. Veh. Technol. **68**(4), 3061–3074 (2019)
18. H. Guo, J. Zhang, J. Liu, FiWi-enhanced vehicular edge computing networks: collaborative task offloading. IEEE Veh. Technol. Mag. **14**(1), 45–53 (2019)
19. F. Sun, F. Hou, N. Cheng, M. Wang, H. Zhou, L. Gui, X. Shen, Cooperative task scheduling for computation offloading in vehicular cloud. IEEE Trans. Veh. Technol. **67**(11), 11049–11061 (2018)
20. Y. Wang, X. Tao, X. Zhang, P. Zhang, Y.T. Hou, Cooperative task offloading in three-tier mobile computing networks: an admm framework. IEEE Trans. Veh. Technol. **68**(3), 2763–2776 (2019)
21. N. Abbas, Y. Zhang, A. Taherkordi, T. Skeie, Mobile edge computing: a survey. IEEE Internet Things J. **5**(1), 450–465 (2018)
22. Y. Mao, C. You, J. Zhang, K. Huang, K.B. Letaief, A survey on mobile edge computing: the communication perspective. IEEE Commun. Surv. Tutor. **19**(4), 2322–2358 (Fourthquarter, 2017)
23. S.N. Shirazi, A. Gouglidis, A. Farshad, D. Hutchison, The extended cloud: review and analysis of mobile edge computing and fog from a security and resilience perspective. IEEE J. Sel. Areas Commun. **35**(11), 2586–2595 (2017)
24. P. Bellavista, S. Chessa, L. Foschini, L. Gioia, M. Girolami, Human-enabled edge computing: exploiting the crowd as a dynamic extension of mobile edge computing. IEEE Commun. Mag. **56**(1), 145–155 (2018)
25. Y. Sun, S. Zhou, J. Xu, EMM: energy-aware mobility management for mobile edge computing in ultra dense networks. IEEE J. Sel. Areas Commun. **35**(11), 2637–2646 (2017)
26. P.S. Bithas, G.P. Efthymoglou, A.G. Kanatas, V2V cooperative relaying communications under interference and outdated CSI. IEEE Trans. Veh. Technol. **67**(4), 3466–3480 (2018)
27. C. Perfecto, J. Del Ser, M. Bennis, Millimeter-wave V2V communications: distributed association and beam alignment. IEEE J. Sel. Areas Commun. **35**(9), 2148–2162 (2017)
28. J. Mei, K. Zheng, L. Zhao, Y. Teng, X. Wang, A latency and reliability guaranteed resource allocation scheme for LTE V2V communication systems. IEEE Trans. Wirel. Commun. **17**(6), 3850–3860 (2018)
29. Y. Hui, Z. Su, T.H. Luan, Collaborative content delivery in software-defined heterogeneous vehicular networks. IEEE/ACM Trans. Netw. **28**(2), 575–587 (2020)
30. A. Ghosh, V.V. Paranthaman, G. Mapp, O. Gemikonakli, J. Loo, Enabling seamless V2I communications: toward developing cooperative automotive applications in vanet systems. IEEE Commun. Mag. **53**(12), 80–86 (2015)
31. J. Chung, M. Kim, Y. Park, M. Choi, S. Lee, H.S. Oh, Time coordinated V2I communications and handover for wave networks. IEEE J. Sel. Areas Commun. **29**(3), 545–558 (2011)
32. Z. Su, Y. Hui, T.H. Luan, S. Guo, Engineering a game theoretic access for urban vehicular networks. IEEE Trans. Veh. Technol. **66**(6), 4602–4615 (2017)
33. Y. Hui, Z. Su, T.H. Luan, J. Cai, Content in motion: an edge computing based relay scheme for content dissemination in urban vehicular networks. IEEE Trans. Intell. Transp. Syst. **20**(8), 3115–3128 (2019)
34. W.L. Tan, W.C. Lau, O. Yue, T.H. Hui, Analytical models and performance evaluation of drive-thru internet systems. IEEE J. Sel. Areas Commun. **29**(1), 207–222 (2011)

35. H. Zhou, B. Liu, F. Hou, T.H. Luan, N. Zhang, L. Gui, Q. Yu, X.S. Shen, Spatial coordinated medium sharing: optimal access control management in drive-thru internet. IEEE Trans. Intell. Transp. Syst. **16**(5), 2673–2686 (2015)
36. J. Du, F.R. Yu, X. Chu, J. Feng, G. Lu, Computation offloading and resource allocation in vehicular networks based on dual-side cost minimization. IEEE Trans. Veh. Technol. **68**(2), 1079–1092 (2019)
37. Q. Qi, J. Wang, Z. Ma, H. Sun, Y. Cao, L. Zhang, J. Liao, Knowledge-driven service offloading decision for vehicular edge computing: a deep reinforcement learning approach. IEEE Trans. Veh. Technol. **68**(5), 4192–4203 (2019)
38. Z. Zhou, H. Liao, X. Zhao, B. Ai, M. Guizani, Reliable task offloading for vehicular fog computing under information asymmetry and information uncertainty. IEEE Trans. Veh. Technol. **68**(9), 8322–8335 (2019)
39. M. Chen, Y. Hao, Task offloading for mobile edge computing in software defined ultra-dense network. IEEE J. Sel. Areas Commun. **36**(3), 587–597 (2018)
40. G. Zhang, F. Shen, Z. Liu, Y. Yang, K. Wang, M. Zhou, FEMTO: fair and energy-minimized task offloading for fog-enabled IoT networks. IEEE Internet Things J. **6**(3), 4388–4400 (2019)
41. Y. Sun, X. Guo, J. Song, S. Zhou, Z. Jiang, X. Liu, Z. Niu, Adaptive learning-based task offloading for vehicular edge computing systems. IEEE Trans. Veh. Technol. **68**(4), 3061–3074 (2019)
42. T.X. Tran, D. Pompili, Joint task offloading and resource allocation for multi-server mobile-edge computing networks. IEEE Trans. Veh. Technol. **68**(1), 856–868 (2019)
43. K. Wang, Y. Tan, Z. Shao, S. Ci, Y. Yang, Learning-based task offloading for delay-sensitive applications in dynamic fog networks. IEEE Trans. Veh. Technol. **68**(11), 11399–11403 (2019)
44. J. Nie, J. Luo, Z. Xiong, D. Niyato, P. Wang, A stackelberg game approach toward socially-aware incentive mechanisms for mobile crowdsensing. IEEE Trans. Wirel. Commun. **18**(1), 724–738 (2019)
45. G. El Rahi, S.R. Etesami, W. Saad, N.B. Mandayam, H.V. Poor, Managing price uncertainty in prosumer-centric energy trading: a prospect-theoretic stackelberg game approach. IEEE Trans. Smart Grid **10**(1), 702–713 (2019)
46. N. Wu, X. Zhou, M. Sun, Secure transmission with guaranteed user satisfaction in hetero-geneous networks: a two-level stackelberg game approach. IEEE Trans. Commun. **66**(6), 2738–2750 (2018)
47. H. Fang, L. Xu, Y. Zou, X. Wang, K.R. Choo, Three-stage stackelberg game for defending against full-duplex active eavesdropping attacks in cooperative communication. IEEE Trans. Veh. Technol. **67**(11), 10788–10799 (2018)
48. L. Shi, L. Zhao, G. Zheng, Z. Han, Y. Ye, Incentive design for cache-enabled D2D underlaid cellular networks using stackelberg game. IEEE Trans. Veh. Technol. **68**(1), 765–779 (2019)
49. Z. Su, Q. Xu, Y. Hui, S. Wen, S. Guo, A game theoretic approach to parked vehicle assisted content delivery in vehicular ad hoc networks. IEEE Trans. Veh. Technol. **66**(7), 6461–6474 (2017)

Chapter 5
Auction Based Secure Computation Offloading in Vehicular Networks

The cloud computing and edge computing technologies have provided new opportunities for vehicular task offloading in vehicular networks. However, the computing task of a vehicle can be offloaded to multiple edge servers. Therefore, how to select the optimal edge server to complete the offloading service is a challenge. In addition, some malicious edge servers may declare unreasonable prices to execute the offloading service. A market mechanism is thus required to constrain the bids of edge servers in the networks. Besides, the edge servers may provide the offloading services with low quality which will decrease the quality of experience (QoE) of the service requesters. In order to solve the above mentioned challenges, this chapter proposes an auction based task offloading framework to ensure that the computing tasks requested by vehicles can be safely offloaded and executed. First, to constrain the bids of the edge servers in the networks, a task offloading scheme based on the first price sealed auction is proposed for the edge servers which intend to join in the task offloading process. Then, considering the service quality of each edge server, the cloud server is used to evaluate the quality of the edge servers. By designing the security evaluation and prediction algorithm for edge servers based on transductive support vector machine (TSVM), the optimal edge server can be selected for the vehicle to execute the offloading task. With the edge-cloud networks, we evaluate the secure offloading strategy of the vehicle, where the simulation results show that the task offloading scheme proposed in this chapter has a higher efficiency than the conventional schemes.

5.1 Introduction

With the rapid development of intelligent transportation system (ITS) and autonomous driving, an increasing number of edge servers are deployed at roadside infrastructures to facilitate various computing applications emerged in vehicular

© Springer Nature Switzerland AG 2021
Z. Su et al., *The Next Generation Vehicular Networks, Modeling, Algorithm, and Applications*, Wireless Networks, https://doi.org/10.1007/978-3-030-56827-6_5

networks [1–5]. With edge computing, a vehicle which intends to complete a computing task can offload the task to the edge server which is close to the vehicle. After the task is completed, the edge server then delivers the task result to the vehicle. In this way, the computing task can be executed by the edge server instead of the cloud server to reduce service time [6–10]. However, due to the vehicular networks are open access networks, there are still many challenges for achieving a secure environment for computation offloading in vehicular networks.

For the edge servers in the vehicular networks, they are deployed by different service providers and have different task execution performances. For example, some malicious edge servers may declare unreasonable prices to execute the offloading service. In addition, some edge servers in the vehicular networks may provide task results with a low quality to obtain the computing reward. Being faced with the above threats, for the vehicles which can connect to more than one edge servers to offload their computing tasks, how to select the optimal edge server to execute the task with the target of minimizing the task execution cost and guaranteeing the quality of the service therefore becomes a challenge. To this end, in this chapter, we propose a secure vehicular task offloading scheme by integrating cloud computing and edge computing. The specific research contents are as follows:

- In order to constrain the bids of edge servers which intend to join in the task offloading process, we develop a task offloading scheme based on the first price sealed auction, where the optimal bidding strategy of each edge server is analyzed.
- Considering the service quality, the cloud server is used to evaluate the service quality of each edge server. By designing the security evaluation algorithm for edge servers based on transductive support vector machine (TSVM), the optimal edge server can be selected for the vehicle to execute the offloading task to minimize the task execution cost and guarantee the quality of the service.

5.2   System Model

In this section, we model the vehicular offloading system in the edge-cloud networks which includes the network model and the task model.

5.2.1   Network Model

Figure 5.1 is a 3-tier hierarchical network of the vehicular offloading scenario in the edge-cloud networks, which mainly includes the vehicle layer, the edge computing layer and the cloud layer [11–15].

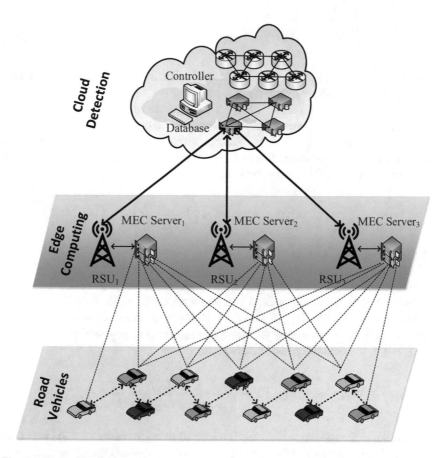

Fig. 5.1 Task offloading model

1. **Vehicle Layer**

   With the continuous enrichment of vehicular applications, vehicles will generate various computing tasks (e.g., route planning and parking space search) and request different entertainment services (e.g., video streaming and augmented reality) during driving [16–19]. According to [20–22], the mean arrival rate of vehicles in the road is mainly related to the speed of the vehicles and the traffic density of the current road conditions, shown as

$$\phi(t) = \rho(t)v(t), \tag{5.1}$$

   where $\rho(t)$ and $v(t)$ represent the vehicle speed and traffic density at the current moment, respectively. Based on [23–25], the relationship between $\rho(t)$ and $v(t)$ is given by

$$v(t) = v_{\max} \left(1 - \frac{\rho(t)}{\rho_{\max}} \right), \tag{5.2}$$

where $\rho_{\max}$ and $v_{\max}$ are the maximum traffic density and the maximum driving speed of the vehicles, respectively.

Let $J = \{1, \ldots, j, \ldots, J\}$ be the mean number of vehicles arriving in the current road area. Assuming that the computing task request generated by each vehicle in the networks obeys a Poisson distribution, where the average arrival rate of the task request is denoted as λ. Therefore, the number of vehicles carrying computing tasks in the current road area is

$$J^*(t) = \lambda v_{\max} \rho(t) \left(1 - \frac{\rho(t)}{\rho_{\max}} \right). \tag{5.3}$$

2. **Edge Computing Layer**
   Compared with the remote cloud server, the edge server is closer to the vehicles, thereby providing the task computing services for the vehicles with low latency. Generally, the edge server is deployed at a roadside unit (RSU) or a base station (BS) [26–30]. Let $I = \{1, \ldots, i, \ldots, I\}$ denote the number of edge servers in the road area. When vehicle j establishes a connection with edge server i, according to [31–35], the data transmission rate is

$$r_{i,j} = W_i \log_2 \left(1 + \frac{p_j h_{i,j}}{\sum_{j' \neq J^*, j' \neq j} p_{j'} h_{i,j'} + \sigma^2} \right), \tag{5.4}$$

where W_i is the spectrum bandwidth allocated to edge server i by the backhaul networks, p_j and $h_{i,j}$ represent the transmission power density of vehicle j and the channel gain between the vehicle and the edge server, respectively. σ^2 represents the power spectral density of the Gaussian white noise.

   With the edge computing, if the vehicle is in the communication coverage of the edge server and decides to offload its computing task to the edge server, a communication connection will be established between them to upload the initial data of the computing task. After completing the computing task, the edge server will return the computing result to the vehicle.

3. **Cloud Layer**
   The cloud server is generally deployed by trusted organizations, such as network operators and government agencies. As shown in Fig. 5.1, all the edge servers in the networks are connected to the cloud server. The database in the cloud server stores the data (i.e., the quality of offloading service) related to the tasks which are executed by each edge server. As an authoritative organization in the networks, the cloud server can fairly evaluate each edge server. Therefore, when a vehicle can offload its computing task to multiple edge servers, the cloud server can detect the available edge servers and select the optimal one for task offloading to ensure the service quality of computing tasks.

5.2.2   Task Model

Due to the diversity of vehicular applications, the computing task requirements generated by different vehicles are also diverse. Let $T_j = \{Z_j, D_j, t_j^{max}\}$ denote the computing task request generated by vehicle j, where the three parameters represent the initial input data size of the computing task, the number of computing resources required to complete the computing task, and the maximum delay to complete the computing task, respectively. Generally, when the input data of a computing task is large, it also requires more computing resources. Therefore, the relationship between D_j and Z_j can be defined by

$$D_j = \zeta Z_j, \tag{5.5}$$

where ζ is a coefficient to describe the characteristics of the computing task.

5.3   Analysis of Secure Offloading Strategy in Edge-Cloud Networks

In this section, the secure offloading strategy of the vehicle in the edge-cloud networks is analyzed. First, based on the consideration of the offloading utility of the vehicle, the first price sealed auction based task offloading scheme is designed. Then, the cloud server (authoritative organization) evaluates the quality of the edge servers according to the TSVM-based security detection algorithm with the target of selecting the optimal edge server to execute the task.

5.3.1   Task Offloading Scheme Based on First Price Sealed Auction

In order to save local computing resources, the vehicle will offload the computing task to the edge server deployed along the roadside [36–40]. After the computing task is executed by the edge server, the computing result will be transmitted to the vehicle. In this subsection, we propose a computation offloading scheme based on the first price sealed auction to constrain the bids of the edge servers. The optimal offloading strategy with two steps is determined for the vehicle by considering the computing capability of each edge server. In the first step, the vehicle selects candidate edge servers according to the requirements of the computing task. In the second step, as a rational individual, the vehicle hopes to minimize the price paid for the task offloading service. Therefore, the first price sealed auction is used to guide the edge servers in the networks to declare the price for the offloading service. In what follows, we detail the process of the offloading scheme.

Step 1: Select Candidate Edge Servers

In the vehicular networks, the computing capability and the resource usage status of edge servers are different. Inevitably, the computing capability of some edge servers at the current time does not satisfy the vehicle's computing task offloading requirements. Therefore, before starting the auction, the computing capability of each edge server needs to be detected to screen out the edge servers that meet the requirements. Specifically, the candidate edge servers are selected based on their computing capability and the threshold set by the vehicle. Let $a_i(t) = 1$ denote that the computing capability of edge server i is no lower than the threshold set by the vehicle at time t, otherwise, $a_i(t) = 0$. We have

$$a_i(t) = \begin{cases} 1, & F_i(t) \geq \varepsilon_j(t), \\ 0, & \text{otherwise}, \end{cases} \tag{5.6}$$

where $F_i(t)$ is the computing capability of edge server i at time t. $\varepsilon_j(t)$ is the threshold set by vehicle j for its computing task. Considering the computing capability of the edge server and the offloading requirements of the computing task, the threshold set by vehicle j can be expressed as

$$\varepsilon_j(t) = \min\{\max\{\Omega_j(t), F_{\min}(t)\}, F_{\max}(t)\}, \tag{5.7}$$

where $F_{\min}(t)$ and $F_{\max}(t)$ are the maximum and minimum computing capabilities of the edge servers at time t, respectively. $\Omega_j(t)$ is a variable controlled by vehicle j, which is used to reflect the computation offloading requirement of the task. It can be expressed as

$$\Omega_j(t) = \frac{F_{\max}(t) + F_{\min}(t)}{\varrho_j} + \tau_j \mu_j(t), \tag{5.8}$$

where $\mu_j(t)$ represents the urgency of the computing task of vehicle j. ϱ_j and τ_j are adjustment factors.

Therefore, for vehicle j, the number of edge servers selected in advance can be expressed as

$$I_j^*(t) = \sum_{i=1}^{I} a_i(t). \tag{5.9}$$

Step 2: Select the Optimal Edge Server

In order to constrain the prices declared by different edge servers, we use the first price sealed auction to select the edge server. The edge server with the lowest bid price wins the auction, thereby providing the offloading service for the vehicle and obtaining the corresponding offloading utility. Through edge computation offloading, vehicles can save local computing resources, computing time, and energy consumption. Therefore, the utility obtained by vehicle j through edge offloading can be expressed as

$$U_j(t) = S_j(f_{i,j}) - b_i, \qquad (5.10)$$

where $S_j(f_{i,j})$ is the satisfaction function of vehicle j for the edge computing offloading. b_i is the bid price of edge server i. According to [41], the satisfaction function of vehicle j can be defined as

$$S_j(f_{i,j}) = w_j \log \left(1 + \frac{f_{i,j}}{f_{i,j}^{\min}} \right). \qquad (5.11)$$

It can be known from (5.11) that the satisfaction of the vehicle is a function of the computing resources allocated by edge server i to vehicle j. $f_{i,j}^{\min}$ is the minimum offloading resources. w_j is the willingness of vehicle j for offloading its computing task, which is expressed by the S-type function, shown as

$$w_j = \frac{\alpha}{1 + e^{-\beta \left(\vartheta - f_j^l \right)}}, \qquad (5.12)$$

where f_j^l is the local computing resources owned by vehicle j. α and β are constant values defined in advance by the vehicle according to the characteristics (such as urgency) of its computing task. ϑ is a fixed value that represents the maximum offloading resource requirement. It can be seen from (5.12) that the offloading willingness of the vehicle decreases with the increase of the local computing resources owned by vehicle j.

From (5.11) and (5.12), the utility function of vehicle j can be rewritten as

$$U_j(t) = \frac{\alpha}{1 + e^{-\beta \left(\vartheta - f_j^l \right)}} \log \left(1 + \frac{f_{i,j}}{f_{i,j}^{\min}} \right) - b_i. \qquad (5.13)$$

In the auction process, the edge server, as a rational bidder, will evaluate the offloading service to obtain the cost spent on completing the task [42–46]. Let c_i denote the task evaluation of edge server i for completing the task. Since the edge server is a rational individual, we have $b_i > c_i$. As such, the utility obtained by edge server i which joins in the auction game is the difference between the bid price and the task valuation. We have

$$U_i(t) = \begin{cases} b_i - c_i, & b_i < \min\{b_{-i}\}, \\ 0, & \text{otherwise}, \end{cases} \qquad (5.14)$$

where b_{-i} represents the bid prices of the edge servers except edge server i. The valuation cost c_i is related to the number of computing resources allocated to the vehicle by the edge server, shown as

$$c_i = \gamma_i f_{i,j}, \qquad (5.15)$$

where γ_i is a coefficient related to the computing resources which is determined by edge server i and obeys a uniform distribution in the range of $[\gamma, \overline{\gamma}]$.

As a rational auction bidder, edge server i needs to determine the optimal bid price in order to maximize the utility obtained from the auction, we have

$$\max_{b_i, f_{i,j}} E\{U_i(t)\}. \tag{5.16}$$

$$s.t. \quad c_i < b_i.$$

We give the following two theorems to solve (5.16).

Theorem 5.1 *For edge server i, the optimal strategy for allocating computing resources to the computing task of vehicle j is*

$$f_{i,j}^* = \arg\max_{f_{i,j}} \left\{ \frac{\alpha}{1 + e^{-\beta\left(\vartheta - f_j^l\right)}} \log\left(1 + \frac{f_{i,j}}{f_{i,j}^{\min}}\right) - \gamma_i f_{i,j} \right\}. \tag{5.17}$$

Proof Assuming that $f_{i,j}^*$ is not the optimal strategy for the computing resources allocated by edge server i to vehicle j, then there must exist another optimal strategy $(\widehat{b}_i, \widehat{f_{i,j}})$, which can maximize the utility obtained by edge server i, where $\widehat{f_{i,j}} \neq f_{i,j}^*$.

Let

$$b_i^* = w_j \log\left(1 + \frac{f_{i,j}^*}{f_{i,j}^{\min}}\right) - \left(w_j \log\left(1 + \frac{\widehat{f_{i,j}}}{f_{i,j}^{\min}}\right) - \widehat{b}_i\right). \tag{5.18}$$

We have

$$U_i\left(\widehat{b}_i, \widehat{f_{i,j}}\right) = \widehat{b}_i - \gamma_i \widehat{f_{i,j}} \tag{5.19}$$

$$= b_i^* - w_j \log\left(1 + \frac{f_{i,j}^*}{f_{i,j}^{\min}}\right) + w_j \log\left(1 + \frac{\widehat{f_{i,j}}}{f_{i,j}^{\min}}\right) - \gamma_i \widehat{f_{i,j}}$$

$$\leq b_i^* - w_j \log\left(1 + \frac{f_{i,j}^*}{f_{i,j}^{\min}}\right) + w_j \log\left(1 + \frac{f_{i,j}^*}{f_{i,j}^{\min}}\right) - \gamma_i f_{i,j}^*$$

$$= U_i\left(b_i^*, f_{i,j}^*\right).$$

Obviously, this is contradictory to the assumption that $(\widehat{b}_i, \widehat{f_{i,j}})$ is the optimal strategy. Thus, $f_{i,j}^*$ is the optimal strategy of the computing resources allocated by edge server i to vehicle j. The theorem is proved.

By solving (5.17), we have

$$f_{i,j}^* = \frac{w_j}{\gamma_i} - f_{i,j}^{\min}. \tag{5.20}$$

Next, we analyze the optimal bid price of edge server i in the auction for completing the computing task requested by vehicle j.

Theorem 5.2 *For edge server i, the optimal strategy for competing with other edge servers is*

$$b_i^* = \gamma_i f_{i,j}^* + \frac{f_{i,j}^*}{I_j^*} (\bar{\gamma} - \gamma_i). \tag{5.21}$$

Proof From (5.14), it can be seen that edge server i can win the auction and obtain the utility if its bid price is the lowest in the auction. Denote by P_i the probability that edge server i wins the auction in the game, then the problem in (5.16) can be rewritten as

$$\max_{b_i, f_{i,j}} E\{U_i(t)\} = \max_{b_i, f_{i,j}} \left\{ P_i(b_i - \gamma_i f_{i,j}) + 0 \cdot (1 - P_i) \right\}. \tag{5.22}$$

Considering the evaluation cost of edge server i during the auction, the maximum offloading utility obtained by vehicle j from edge server i can be given by

$$\Phi_{i,j} = w_j \log \left(1 + \frac{f_{i,j}^*}{f_{i,j}^{\min}} \right) - \gamma_i f_{i,j}^*. \tag{5.23}$$

On the other hand, when the computing resources allocated by edge server i for the computing task is determined, the benefit obtained by vehicle j is

$$\tilde{U}_{i,j} = w_j \log \left(1 + \frac{f_{i,j}^*}{f_{i,j}^{\min}} \right) - b_i. \tag{5.24}$$

By substituting (5.23) and (5.24) into (5.22), it can be obtained that the expected utility of edge server i is

$$E\{U_i(t)\} = \left(\Phi_{i,j} - \tilde{U}_{i,j} \right) P_i. \tag{5.25}$$

For vehicle j, its goal is to select an optimal edge server from the candidate edge servers to maximize the utility, we have

$$U_j^*(t) = \max \left\{ \tilde{U}_{i,j} | i = 1, 2, \ldots, I_j^*(t) \right\}. \tag{5.26}$$

Therefore, for each edge server, an optimal bid strategy will be selected to maximize the utility of vehicle j in (5.24).

By considering the maximum utility of vehicle j evaluated by the edge server, the reverse auction is used to obtain the optimal bid price of edge server i. Let $\widetilde{U}_{i,j} = \Psi(\Phi_{i,j})$ denote the bid strategy with which edge server i can win the auction, where the value of $\widetilde{U}_{i,j}$ increases with the increase of $\Phi_{i,j}$. Based on the auction rules, edge server i wins the auction if its bid price is the lowest one. Otherwise, it loses the chance to execute the computing task and its utility is zero. Therefore, we have

$$P_i = \prod_{i'=1, i' \neq i}^{I_j^*} Prob\left\{\widetilde{U}_{i,j} \geq \widetilde{U}_{i',j}\right\}. \tag{5.27}$$

As $\widetilde{U}_{i,j} = \Psi(\Phi_{i,j})$, we have

$$Prob\left\{\widetilde{U}_{i,j} \geq \widetilde{U}_{i',j}\right\} = Prob\left\{\Phi_{i,j} \geq \Phi_{i',j}\right\}. \tag{5.28}$$

In the auction, the edge server does not know the bid strategies of other edge servers. This is because different edge servers have different valuations for the computing task requested by vehicle j. We have

$$H(\Phi_{i,j}) = Prob\left\{\Phi_{i,j} \geq \Phi_{i',j}\right\} = Prob\{\gamma_i \leq \gamma_{i'}\} = 1 - G(\gamma_i), \tag{5.29}$$

where $G(\gamma_i)$ is the probability distribution function of $\gamma_{i'}$. By substituting (5.29) into (5.27), we have

$$P_i = H(\Phi_{i,j})^{I_j^*-1} = (1 - G(\gamma_i))^{I_j^*-1}. \tag{5.30}$$

Let $F(\Phi_{i,j}) = P_i$, the utility obtained by edge server i can be rewritten as

$$E\{U_i(t)\} = (\Phi_{i,j} - \widetilde{U}_{i,j})F(\Phi_{i,j}). \tag{5.31}$$

To obtain the maximum value of (5.31), we calculate the first derivative of $E\{U_i(t)\}$ with respect to $\widetilde{U}_{i,j}$, shown as

$$\frac{\partial E\{U_i(t)\}}{\partial \widetilde{U}_{i,j}} = \frac{(\Phi_{i,j} - \widetilde{U}_{i,j})(\widehat{F}(\Psi^{-1}(\widetilde{U}_{i,j})))}{\Psi'(\Psi^{-1}(\widetilde{U}_{i,j}))} - F(\Psi^{-1}(\widetilde{U}_{i,j})). \tag{5.32}$$

Let $\frac{\partial E\{U_i(t)\}}{\partial \widetilde{U}_{i,j}}$ equal to 0, we have

$$\frac{\partial(F(\Phi_{i,j})\Psi(\Phi_{i,j}))}{\partial \Phi_{i,j}} = \widehat{F}(\Phi_{i,j})\Phi_{i,j}. \tag{5.33}$$

Therefore,

$$\tilde{U}_{i,j} = \Psi(\Phi_{i,j}) = \frac{1}{F(\Phi_{i,j})} \int_0^{\Phi_{i,j}} \widehat{F}(x)x dx = \Phi_{i,j} - \frac{1}{F(\Phi_{i,j})} \int_0^{\Phi_{i,j}} \widehat{F}(x) dx. \tag{5.34}$$

According to (5.23) and (5.24), the optimal bid price of edge server i in the auction process is

$$b_i^* = \Phi_{i,j} + \gamma_i f_{i,j}^* - \tilde{U}_{i,j}. \tag{5.35}$$

Combining (5.34) and (5.35), we have

$$b_i^* = \gamma_i f_{i,j}^* + \frac{1}{F(\Phi_{i,j})} \int_0^{\Phi_{i,j}} \widehat{F}(x) dx. \tag{5.36}$$

As the parameter γ_i obeys a uniform distribution in the range of $[\underline{\gamma}, \overline{\gamma}]$, its probability distribution function can be expressed as

$$\overline{F}(x) = \frac{x - \underline{\gamma}}{\overline{\gamma} - \underline{\gamma}}. \tag{5.37}$$

Therefore, the optimal bid price of edge server i in (5.36) is

$$b_i^* = \gamma_i f_{i,j}^* + \frac{f_{i,j}^*}{I_j^*}(\overline{\gamma} - \gamma_i). \tag{5.38}$$

The theorem is proved.

After the edge servers in the auction have determined their optimal bid prices $\left(\text{i.e., } \{b_1^*, \ldots, b_i^*, \ldots, b_{I_j^*}^*\}\right)$, vehicle j will select the edge server with the lowest bid price to complete the offloading task, shown as

$$i^* = \arg\min b_i^*, i \in I = \left\{1, \ldots, i, \ldots, I_j^*\right\}. \tag{5.39}$$

According to the above analysis, the detailed process of selecting the edge server based on the first price sealed auction is shown in Algorithm 5.1.

In addition to the bid prices of the edge servers, the quality of the computation offloading service needs to be judged to select the edge server to execute the offloading task. This is because in the auction process, some edge servers with low service quality will maliciously reduce the bid price to win the auction, which will bring a low QoE to the vehicle. Therefore, in the auction process of the computing task, the bid prices of the edge servers are sorted in an ascending order, denoted as

---

Algorithm 5.1 The first price sealed auction based edge server selection scheme

---

1: **Input:** The task $T_j = \{Z_j, D_j, t_j^{\max}\}$, the maximum computing capability $F_i^{\max}$, the minimum computing demand $f_{i,j}^{\min}$

2: **Output:** The optimal edge server i^*, the optimal resource demand $f_{i,j}^*$

3: **Stage 1: Select the candidate edge servers**

4: Set $I_j^*(t) = 0$

5: Vehicle j calculates $\varepsilon_j(t)$ by (5.7) and declares it to edge servers

6: **for** $i = 1, i \leq I_j(t)$ **do**

7:     **if** Decide to join in the auction **then**

8:         **if** $a_i(t) = 1$ **then**

9:             $I_j^*(t) = I_j^*(t) + 1$

10:         **end if**

11:     **end if**

12: **end for**

13: **Stage 2: Select the optimal edge server for offloading the task**

14: **for** $i = 1, i \leq I_j^*(t)$ **do**

15:     Calculate the optimal resource demand $f_{i,j}^*$ by (5.20)

16:     **if** $f_{i,j}^* \leq f_{i,j}^{\min} \& f_{i,j}^{\min} \leq F_i^{\max}$ **then**

17:         $f_{i,j}^* = f_{i,j}^{\min}$

18:     **else**

19:         **if** $f_{i,j}^{\min} > F_i^{\max}$ **then**

20:             $f_{i,j}^* = 0$

21:         **else**

22:             **if** $f_{i,j}^* > F_i^{\max}$ **then**

23:                 $f_{i,j}^* = F_i^{\max}$

24:             **end if**

25:         **end if**

26:     **end if**

27:     Calculate the optimal bid price b_i^* by (5.21)

28: **end for**

29: Select the edge server by (5.39)

$\left\{b_1', \ldots, b_i', \ldots, b_{I_j^*}'\right\}$. Then, the security of the edge servers is detected based on the sorted bid prices to determine the service quality provided by the corresponding edge server. In this way, the optimal edge server can be determined to provide efficient and reliable computation offloading services for vehicles.

5.3.2   TSVM-Based Detection Scheme

According to the auction based computation offloading scheme, the optimal bid prices of all edge servers in the auction can be obtained. In order to maximize the utility of the vehicle, the edge server with the lowest bid price will be the winner of the auction. However, not all the edge servers are willing to provide offloading services with high quality. Some edge servers with low service quality may reduce their bid prices maliciously in order to win the auction. As such, if the computing task of the vehicle is offloaded to a malicious edge server, the quality of computing service will be reduced. To solve this problem, a security detection scheme based on TSVM is proposed to assess the quality of edge servers. In the edge-cloud networks, the cloud server acts as a trusted authority. By connecting the edge server, the cloud server's database stores all the data of computing services provided by the edge servers, including the characteristics of the computing tasks, execution time, quality of the offloading service, etc. Using the historical data of an edge server, the cloud server can predict whether the edge server is reliable or not. In this way, the cloud server can assist the vehicle to select the optimal edge server for computation offloading.

Let $A_l = \{(x_1, y_1), (x_2, y_2), \ldots, (x_l, y_l)\}$ denote the set of labeled samples about edge servers in the database of the cloud server. Here, x_i is a feature vector of edge server i. It represents the parameters of computing tasks and the execution time. $y_i \in \{-1, 1\}$ denotes the evaluation of the offloading quality, which is evaluated by the cloud server. Specifically, $y_i = 1$ means that the quality of offloading services provided by this edge server is acceptable. In addition, there are a large number of unlabeled samples in the database, denoted by $A_u = \{x_{l+1}, x_{l+2}, \ldots, x_{l+u}\}$. We use TSVM to predict and classify the offloading quality of the edge server. The learning goal is to give the prediction results $\widehat{y} = (\widehat{y}_{l+1}, \widehat{y}_{l+2}, \ldots, \widehat{y}_{l+u})$ for the samples in A_u. We have

$$\min \frac{1}{2}\|w\|_2^2 + B_l \sum_{i=1}^{l} \xi_i + B_u \sum_{i=l+1}^{l+u} \xi_i, \tag{5.40}$$

$$s.t. \quad y_i\left(w^T x_i + b\right) \geq 1 - \xi_i, i = 1, 2, \ldots, l,$$

$$\widehat{y}_i\left((w^T x_i + b\right) \geq 1 - \xi_i, i = l+1, l+2, \ldots, l+u,$$

$$\xi_i \geq 0, i = 1, 2, \ldots, l+u.$$

where (w, b) is the hyperplane. ξ_i is the slack variable. B_l and B_u are weight parameters determined by the cloud server to balance the importance of labeled and unlabeled samples.

The detailed process of the quality prediction algorithm based on TSVM is shown in Algorithm 5.2. Based on the prediction of Algorithm 5.2, the cloud server can perform an accurate quality evaluation of offloading service for the edge server. In this way, the cloud server evaluates each edge server based on their bids. Specifically, if the service quality of the edge server which bids with b'_1 is unacceptable (e.g., $\widehat{y}'_1 = -1$), the next edge server will be evaluated until $\widehat{y}'_i = 1$. As a result, the edge server which leads to $\widehat{y}'_i = 1$ is selected to execute the offloading task.

Algorithm 5.2 The TSVM based security prediction and evaluation
---
1: **Input:** The labeled samples $A_l = \{(x_1, y_1), (x_2, y_2), \ldots, (x_l, y_l)\}$, the unlabeled samples
$A_u = \{x_{l+1}, x_{l+2}, \ldots, x_{l+u}\}$, the parameters B_l, B_u
2: **Output:** The prediction results $\widehat{y} = (\widehat{y}_{l+1}, \widehat{y}_{l+2}, \ldots, \widehat{y}_{l+u})$
3: Train a SVM_l model by A_l
4: Predict A_u by SVM_l
5: **Initialize:** $B_u \ll B_l$
6: **while** $B_u < B_l$ **do**
7:     Calculate (w, b) and ξ_i by solving (5.40) based on $A_l, A_u, \widehat{y}, B_l, B_u$
8:     **while** $\exists \{i, j | (\widehat{y}_i \widehat{y}_j < 0) \wedge (\xi_i > 0) \wedge (\xi_j > 0) \wedge (\xi_i + \xi_j > 2)\}$ **do**
9:         $\widehat{y}_i = -\widehat{y}_i$
10:        $\widehat{y}_j = -\widehat{y}_j$
11:        Calculate (w, b) and ξ_i by solving (5.40) based on $A_l, A_u, \widehat{y}, B_l, B_u$
12:    **end while**
13:    $B_u = \min\{2B_u, B_l\}$
14: **end while**

5.4 Simulation

This section evaluates the performance of the proposed auction based secure computation offloading scheme by simulations. We first introduce the simulation experiment setting, followed by the analysis of the simulation results.

5.4.1 Setting

In the simulation, the willingness of the vehicle to offload its task changes from 20% to 100%. In the proposed auction based edge server selection scheme, the number of edge servers participating in the auction is selected from $\{2, 4, 6, 8, 10\}$. The minimum computing resources allocated by each edge server is set to be 0.1. The values of $\underline{\gamma}$ and $\overline{\gamma}$ are set to be 0 and 1, respectively.

We evaluate the performance of the proposed edge server selection scheme by comparing with two traditional schemes shown as follows.

- **Random scheme**: The vehicle randomly selects an edge server to offload its computing task.
- **Greedy scheme**: The vehicle offloads its computing task to the edge server which bids with the lowest for providing the computing resources.

5.4.2 Results Analysis

Figure 5.2 depicts the optimal computing resources allocated by the edge server to execute the computing task by changing the willingness of the vehicle. From this

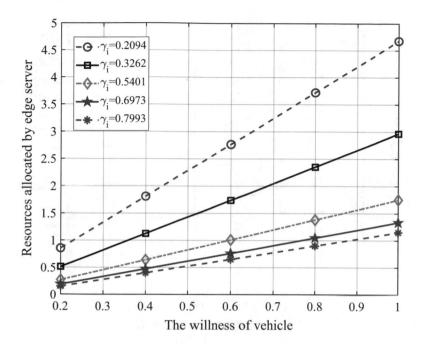

Fig. 5.2 The optimal computing resources allocated by the edge server to execute the computing task by changing the willingness of the vehicle

figure, we can see that the computing resources allocated by the edge server increase with the increase of the willingness of the vehicle. If the willingness of the vehicle is low, the edge server will turn to other vehicles which have a high willingness to allocate computing resources to earn more utility. In addition, we can see that the computing resources allocated by the edge server decrease with the increase of γ_i. This is because that with the increase of γ_i, the cost for the computing resources becomes high. As a result, the computing resources allocated to this vehicle will be decreased.

Figure 5.3 shows the relationship between the optimal bid price of the edge server to win the auction and the number of edge servers participating in the game. Obviously, the bid price of the edge server that wins the auction gradually decreases with the increase of the number of edge servers in the auction game. This is because the more edge servers in the auction, the more fierce competition among the edge servers to win the auction. Consequently, the declared bid price of each edge server becomes low.

Figure 5.4 shows the utility of the vehicle by changing the willingness of the vehicle to offload the computing task. From this figure, we can see that the proposed scheme can lead to the highest utility to the vehicle. In the random scheme, the vehicle randomly selects the edge server for offloading. As a result, if the edge server allocates fewer computing resources with a higher price, the utility of the vehicle will be lower. In addition, in the greedy scheme, the vehicle will always

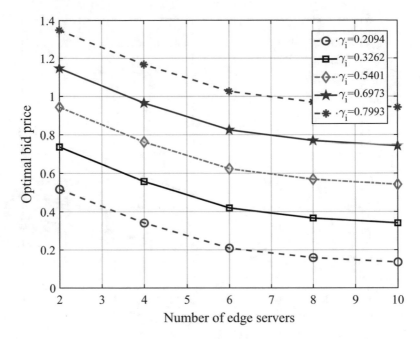

Fig. 5.3 The relationship between the optimal bid price of the edge server to win the auction and the number of edge servers

select the edge server with the lowest offloading price to minimize the cost. As a rational individual, the edge server has fewer computing resources allocated to the vehicle when the offloading price is lower. As such, the utility of the vehicle is low. Different from these two schemes, the proposed scheme enables edge servers to participate in the auction to compete with each other. With the optimal bid price and computing resources allocated to the task, the vehicle can obtain the highest utility.

5.5   Summary

This chapter has presented an auction based secure task offloading scheme in edge-cloud networks. In the proposed scheme, the task offloaded by a vehicle is executed by the edge server and the service quality of each edge server is evaluated by the cloud server. Specifically, we have proposed a task offloading scheme based on the first price sealed auction to constrain the bid prices of edge servers in the vehicular networks. By sorting the bid prices of the edge servers, a TSVM based service quality evaluation and prediction algorithm has been designed to detect the service quality of each edge server. With the auction game and the quality prediction algorithm, the optimal edge server can be selected to execute the offloaded computing task. The simulation results have shown that the proposed edge server selection scheme can lead to the highest utility to the vehicle.

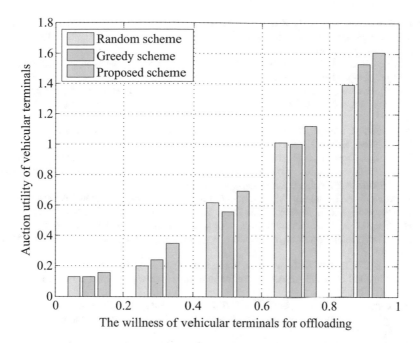

Fig. 5.4 The utility of the vehicle by changing the willingness of the vehicle to offload the computing task

References

1. Q. Yuan, H. Zhou, J. Li, Z. Liu, F. Yang, X.S. Shen, Toward efficient content delivery for automated driving services: an edge computing solution. IEEE Netw. **32**(1), 80–86 (2018)
2. Z. Su, Y. Hui, T.H. Luan, Distributed task allocation to enable collaborative autonomous driving with network softwarization. IEEE J. Sel. Areas Commun. **36**(10), 2175–2189 (2018)
3. J. Zhang, K.B. Letaief, Mobile edge intelligence and computing for the internet of vehicles. Proc. IEEE **108**(2), 246–261 (2020)
4. Y. Hui, Z. Su, T.H. Luan, J. Cai, Content in motion: an edge computing based relay scheme for content dissemination in urban vehicular networks. IEEE Trans. Intell. Transp. Syst. **20**(8), 3115–3128 (2019)
5. S. Liu, L. Liu, J. Tang, B. Yu, Y. Wang, W. Shi, Edge computing for autonomous driving: opportunities and challenges. Proc. IEEE **107**(8), 1697–1716 (2019)
6. J. Zhang, H. Guo, J. Liu, Y. Zhang, Task offloading in vehicular edge computing networks: a load-balancing solution. IEEE Trans. Veh. Technol. **69**(2), 2092–2104 (2020)
7. Y. Liu, H. Yu, S. Xie, Y. Zhang, Deep reinforcement learning for offloading and resource allocation in vehicle edge computing and networks. IEEE Trans. Veh. Technol. **68**(11), 11158–11168 (2019)
8. H. Guo, J. Zhang, J. Liu, FiWi-enhanced vehicular edge computing networks: collaborative task offloading. IEEE Vehic. Technol. Mag. **14**(1), 45–53 (2019)
9. Z. Su, Y. Hui, Q. Xu, T. Yang, J. Liu, Y. Jia, An edge caching scheme to distribute content in vehicular networks. IEEE Trans. Veh. Technol. **67**(6), 5346–5356 (2018)

10. L. Pu, X. Chen, G. Mao, Q. Xie, J. Xu, Chimera: an energy-efficient and deadline-aware hybrid edge computing framework for vehicular crowdsensing applications. IEEE Internet Things J. **6**(1), 84–99 (2019)
11. W. Chen, B. Liu, H. Huang, S. Guo, Z. Zheng, When UAV swarm meets edge-cloud computing: the QoS perspective. IEEE Netw. **33**(2), 36–43 (2019)
12. J. Guo, B. Song, S. Chen, F.R. Yu, X. Du, M. Guizani, Context-aware object detection for vehicular networks based on edge-cloud cooperation. IEEE Internet Things J. **PP**(99), 1–9 (2019)
13. A. Ceselli, M. Premoli, S. Secci, Mobile edge cloud network design optimization. IEEE/ACM Trans. Netw. **25**(3), 1818–1831 (2017)
14. K. Kaur, S. Garg, G. Kaddoum, S.H. Ahmed, F. Gagnon, M. Atiquzzaman, Demand-response management using a fleet of electric vehicles: an opportunistic-SDN-based edge-cloud framework for smart grids. IEEE Netw. **33**(5), 46–53 (2019)
15. A. Aissioui, A. Ksentini, A.M. Gueroui, T. Taleb, On enabling 5G automotive systems using follow me edge-cloud concept. IEEE Trans. Veh. Technol. **67**(6), 5302–5316 (2018)
16. H. Peng, Q. Ye, X.S. Shen, SDN-based resource management for autonomous vehicular networks: a multi-access edge computing approach. IEEE Wirel. Commun. **26**(4), 156–162 (2019)
17. C. Yu, B. Lin, P. Guo, W. Zhang, S. Li, R. He, Deployment and dimensioning of fog computing-based internet of vehicle infrastructure for autonomous driving. IEEE Internet Things J. **6**(1), 149–160 (2019)
18. A. Ndikumana, N.H. Tran, D.H. Kim, K.T. Kim, C.S. Hong, Deep learning based caching for self-driving cars in multi-access edge computing. IEEE Trans. Intell. Transp. Syst. **PP**(99), 1–16 (2020)
19. S. Liu, L. Liu, J. Tang, B. Yu, Y. Wang, W. Shi, Edge computing for autonomous driving: opportunities and challenges. Proc. IEEE **107**(8), 1697–1716 (2019)
20. Z. Su, Y. Hui, T.H. Luan, S. Guo, Engineering a game theoretic access for urban vehicular networks. IEEE Trans. Veh. Technol. **66**(6), 4602–4615 (2017)
21. W.L. Tan, W.C. Lau, O. Yue, T.H. Hui, Analytical models and performance evaluation of drive-thru internet systems. IEEE J. Sel. Areas Commun. **29**(1), 207–222 (2011)
22. H. Zhou, B. Liu, T.H. Luan, F. Hou, L. Gui, Y. Li, Q. Yu, X. Shen, Chaincluster: engineering a cooperative content distribution framework for highway vehicular communications. IEEE Trans. Intell. Transp. Syst. **15**(6), 2644–2657 (2014)
23. Y. Hui, Z. Su, S. Guo, Utility based data computing scheme to provide sensing service in internet of things. IEEE Trans. Emerg. Top. Comput. **7**(2), 337–348 (2019)
24. J. Harri, F. Filali, C. Bonnet, Mobility models for vehicular ad hoc networks: a survey and taxonomy. IEEE Commun. Surv. Tutor. **11**(4), 19–41 (2009)
25. Y. Hui, Z. Su, T.H. Luan, J. Cai, A game theoretic scheme for optimal access control in heterogeneous vehicular networks. IEEE Trans. Intell. Transp. Syst. **20**(12), 4590–4603 (2019)
26. H. Liu, Y. Zhang, T. Yang, Blockchain-enabled security in electric vehicles cloud and edge computing. IEEE Netw. **32**(3), 78–83 (2018)
27. L. Zhou, L. Yu, S. Du, H. Zhu, C. Chen, Achieving differentially private location privacy in edge-assistant connected vehicles. IEEE Internet Things J. **6**(3), 4472–4481 (2019)
28. G. Sun, F. Zhang, D. Liao, H. Yu, X. Du, M. Guizani, Optimal energy trading for plug-in hybrid electric vehicles based on fog computing. IEEE Internet Things J. **6**(2), 2309–2324 (2019)
29. N. Cheng, F. Lyu, W. Quan, C. Zhou, H. He, W. Shi, X. Shen, Space/aerial-assisted computing offloading for IoT applications: a learning-based approach. IEEE J. Sel. Areas Commun. **37**(5), 1117–1129 (2019)
30. J. Kang, R. Yu, X. Huang, M. Wu, S. Maharjan, S. Xie, Y. Zhang, Blockchain for secure and efficient data sharing in vehicular edge computing and networks. IEEE Internet Things J. **6**(3), 4660–4670 (2019)
31. Q. Zheng, K. Zheng, L. Sun, V.C.M. Leung, Dynamic performance analysis of uplink transmission in cluster-based heterogeneous vehicular networks. IEEE Trans. Veh. Technol. **64**(12), 5584–5595 (2015)

32. M. Patra, R. Thakur, C.S.R. Murthy, Improving delay and energy efficiency of vehicular networks using mobile femto access points. IEEE Trans. Veh. Technol. **66**(2), 1496–1505 (2017)
33. Y. Ruan, Y. Li, C. Wang, R. Zhang, H. Zhang, Power allocation in cognitive satellite-vehicular networks from energy-spectral efficiency tradeoff perspective. IEEE Trans. Cogn. Commun. Netw. **5**(2), 318–329 (2019)
34. K. Xiong, Y. Zhang, P. Fan, H. Yang, X. Zhou, Mobile service amount based link scheduling for high-mobility cooperative vehicular networks. IEEE Trans. Veh. Technol. **66**(10), 9521–9533 (2017)
35. Y. Tang, N. Cheng, W. Wu, M. Wang, Y. Dai, X. Shen, Delay-minimization routing for heterogeneous vanets with machine learning based mobility prediction. IEEE Trans. Veh. Technol. **68**(4), 3967–3979 (2019)
36. F. Guo, Z. Wang, S. Du, H. Li, H. Zhu, Q. Pei, Z. Cao, J. Zhao, Detecting vehicle anomaly in the edge via sensor consistency and frequency characteristic. IEEE Trans. Veh. Technol. **68**(6), 5618–5628 (2019)
37. Y. Wang, P. Lang, D. Tian, J. Zhou, X. Duan, Y. Cao, D. Zhao, A game-based computation offloading method in vehicular multi-access edge computing networks. IEEE Internet Things J. **PP**(99), 1–11 (2020)
38. Z. Zhou, B. Wang, M. Dong, K. Ota, Secure and efficient vehicle-to-grid energy trading in cyber physical systems: integration of blockchain and edge computing. IEEE Trans. Syst. Man Cyb. Syst. **50**(1), 43–57 (2020)
39. L.T. Tan, R.Q. Hu, Mobility-aware edge caching and computing in vehicle networks: a deep reinforcement learning. IEEE Trans. Veh. Technol. **67**(11), 10190–10203 (2018)
40. J. Liu, H. Guo, J. Xiong, N. Kato, J. Zhang, Y. Zhang, Smart and resilient EV charging in SDN-enhanced vehicular edge computing networks. IEEE J. Sel. Areas Commun. **38**(1), 217–228 (2020)
41. Q. Xu, Z. Su, Q. Zheng, M. Luo, B. Dong, Secure content delivery with edge nodes to save caching resources for mobile users in green cities. IEEE Trans. Ind. Inform. **14**(6), 2550–2559 (2018)
42. J. Kim, J. Lee, S. Park, J.K. Choi, Battery-wear-model-based energy trading in electric vehicles: a naive auction model and a market analysis. IEEE Trans. Ind. Inf. **15**(7), 4140–4151 (2019)
43. J.J.Q. Yu, A.Y.S. Lam, Z. Lu, Double auction-based pricing mechanism for autonomous vehicle public transportation system. IEEE Trans. Intell. Veh. **3**(2), 151–162 (2018)
44. A. Yassine, M.S. Hossain, G. Muhammad, M. Guizani, Double auction mechanisms for dynamic autonomous electric vehicles energy trading. IEEE Trans. Veh. Technol. **68**(8), 7466–7476 (2019)
45. D. Li, Q. Yang, D. An, W. Yu, X. Yang, X. Fu, On location privacy-preserving online double auction for electric vehicles in microgrids. IEEE Internet Things J. **6**(4), 5902–5915 (2019)
46. M. Zeng, S. Leng, S. Maharjan, S. Gjessing, J. He, An incentivized auction-based group-selling approach for demand response management in V2G systems. IEEE Trans. Ind. Inf. **11**(6), 1554–1563 (2015)

Chapter 6
Bargain Game Based Secure Content Delivery in Vehicular Networks

Vehicular networks become essential in providing safety driving services for drivers. With the popularization of the vehicular networks, to protect the security of content delivery becomes a challenge. To this end, in this chapter, we propose a secure content delivery mechanism in the network. Specifically, we first evaluate the trust values of vehicles and roadside units (RSUs) by introducing authority units (AU) to limit the actions of nodes in the networks. Next, a price competitive scheme between the vehicle and the RSU is proposed by using the bargain game to encourage them to improve their utilities and trust. With the designed bargain game based scheme, the security of nodes and packets can be guaranteed. Finally, we evaluate the proposed scheme using simulation. The result demonstrates that the proposed scheme can increase the utilities of vehicles and RSUs to deliver content securely compared to the conventional method.

6.1 Introduction

As one of the key points in smart cities, the transportation issues have been studied in a number of researches in recent years. Along with this, a large amount of information including safety and entertainment information needs to be shared among vehicles [1–4]. To this end, vehicular networks [5–9] become an important role in providing a large number of safety and non-safety applications such as security alarms in emergency situations, entertainment applications, and so on. In addition, governments and private agencies around the world have invested a lot of money in a number of different projects to improve the safety of vehicles and pedestrians on the road. In vehicular networks, a vehicle can be regarded as a node, and vehicles on the road can communicate with each other through point-to-point (P2P) methods or by communicating with the infrastructures deployed along the roadside. The former is called V2V (vehicle-vehicle) communications [10–12],

© Springer Nature Switzerland AG 2021
Z. Su et al., *The Next Generation Vehicular Networks, Modeling, Algorithm, and Applications*, Wireless Networks, https://doi.org/10.1007/978-3-030-56827-6_6

and the latter is called V2I (vehicle-infrastructure) communications [13–16]. The infrastructures generally refer to roadside units (RSUs) placed on both sides of the road, which can act as smart routers to control all vehicles' activities. The goal of traditional vehicular networks is to enable real-time communication between vehicles with or without the help of RSUs through the advanced wireless access technologies, thereby improving the traffic safety and efficiency.

As vehicles in the road connecting to the Internet more and more frequently, network congestion becomes more serious than before [17]. The heavy load of the networks brings a remarkable delay for downloading contents from service providers. The corresponding costs (e.g., transmission cost) for the long time connection becomes high for both vehicles and service providers. To address this problem, RSUs are placed along the roadside to provide network connection. With the RSUs, some contents in the networks can be cached in advance. By doing this, vehicles can download the requested content from its connected RSU, where the data traffic in the networks can be reduced [18]. On the other hand, the delay to download the requested contents becomes shorter for vehicles since vehicles can obtain such contents from local RSUs rather than remote cloud server [19–22].

The typical caching mechanism in vehicular networks is mainly composed of a three-level structure as shown in Fig. 6.1 [23–27]. The first level is the network gateway caches contents from the cloud, the second level is the RSU caches contents from the network gateway, the third level is the vehicles on the road cache contents from the RSUs. If a vehicle requires content from its connected RSU, the RSU checks its cache first, if it has cached the requested content, the content will be

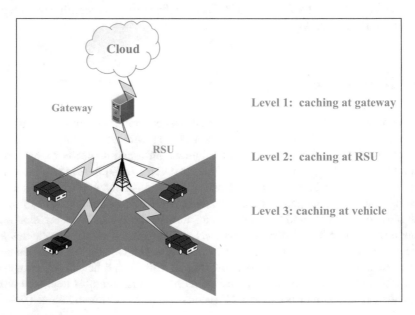

Fig. 6.1 Three levels of caching architecture in vehicular networks

delivered to the vehicle directly through V2I communication. Otherwise, the RSU will ask the network gateway to request the corresponding content.

Although the RSUs with cache can effectively reduce the content downloading delay, the caching system in the networks faces the following new challenges [28–30]. In vehicular networks, the traffic in each area is not evenly distributed. If all vehicles only select the RSU according to the nearest selection method, it will cause problems such as excessive load on some RSUs. Meanwhile, some RSUs keep a lot of idle resources, resulting in the waste of resources. In addition, RSUs and vehicles in the networks may be hacked. Specifically, when a vehicle communicates with the connected RSU or other vehicles, the vehicle may need to send their private contents (e.g., location and hobby), which may be attacked by the malicious nodes [31]. On the other hand, the quality and accuracy of the content provided by the RSU need to be further studied. This is because the fact that the RSUs or the vehicles in the networks may have malicious behaviors when participating in interaction activities. However, most of the existing security solutions focus on the cryptographic method (e.g., signatures, certificates and Public Key Infrastructures (PKIs) [32]) to deal with the outside unauthorized attacks, where the performance is not satisfied to resolve the inside attacks (e.g., hijacking, imitating legitimate vehicles, etc.).

Therefore, this chapter studies the security issues when vehicles communicate with RSUs by proposing a secure content delivery scheme in vehicular networks. By considering the trustworthiness of nodes, the vehicle can choose the appropriate RSU for securing the communication, where the security of connection can be ensured. Firstly, with authority units (AU), a trust evaluation mechanism is introduced for all the nodes in the network. Secondly, based on the trust evaluation mechanism, a bargain game model is presented for the nodes to enhance their trust values and improve their utilities. With the proposed game model, both the vehicle and the RSU can decide the optimal price to make connection with each other with the goal of achieving high utilities. Finally, the simulation experiment is carried out to evaluate the performance of the proposal.

6.2   Problem Formulation

In this section, we first define the traffic load by considering the load balance in the networks. Then, the solution for deploying RSUs is discussed by defining the load of each RSU. After that, the trust values of vehicles and RSUs are evaluated separately by introducing the AU. Among them, the trust values of vehicle users are determined by the evaluation of an AU, and the trust values of RSU are divided into direct trust evaluation and indirect trust evaluation. In order to quickly and fully evaluate the trust values of the nodes in the network, we consider the ratio of the traffic load to the total load of each small area in the network to give an optimization deployment plan for the AU vehicles in each small area.

6.2.1 Deployment of RSUs

In vehicular networks, due to the dynamic traffic, the traffic load in different locations is different from each other. Therefore, if each vehicle user only sends a service request to the RSU in the current area, the load of the RSU in a dense area will be oversaturated, resulting in failure to deliver the requested content to the vehicle in time. In the area with sparse traffic, since RSU can hardly receive any requests, it will cause a lot of waste of resources. Although we can simply deploy more RSUs in the areas to solve this problem, RSUs are expensive and require significant maintenance costs. From an economic point of view, this method is obviously not very economical. Therefore, in order to deploy RSUs more economically, in this subsection, we consider to specifically define the traffic load of each place in vehicular networks, and then design a method for deploying RSUs in some areas with a high load, as shown in Fig. 6.2. By defining the load of each RSU and updating the load of the RSU on the cloud platform in real time, the vehicle user can no longer turn to the RSU that is closest to the vehicle when it needs to send a query request, but comprehensively considers the load of each RSU, the distance from these RSUs and other factors to determine which RSU can be selected to complete the content delivery service. In this way, the load of each RSU in the vehicular networks can be balanced, where the query satisfaction rate of vehicle users is thus increased.

The load of the networks and the deployment of the RSUs are defined as follows. The area of vehicular networks L is divided into several subareas $L = \{1, \ldots, l, \ldots, L\}$. The travelling time of vehicle user $v \in \{1, \ldots, v, \ldots, V\}$ from subarea i to j is denoted as $t_v(i, j)$. Therefore, we can define the connectivity of i and j as $C_{ij} = \sum_{v=1}^{V} t_v(i, j)$. In this way, we can define the connectivity metric of

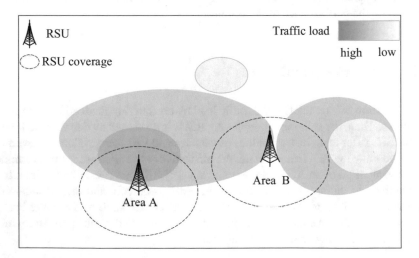

Fig. 6.2 Deployment of RSUs in different areas

all subareas as

$$
C = \begin{bmatrix}
0 & c_{12} & c_{13} & \cdots & c_{1l} & \cdots & c_{1L} \\
c_{21} & 0 & c_{23} & \cdots & c_{2l} & \cdots & c_{2L} \\
c_{31} & c_{32} & 0 & \cdots & c_{3l} & \cdots & c_{3L} \\
\vdots & \vdots & \vdots & \ddots & \vdots & \vdots & \vdots \\
c_{l1} & c_{l2} & c_{l3} & \cdots & 0 & \cdots & c_{lL} \\
\vdots & \vdots & \vdots & \vdots & \vdots & 0 & \vdots \\
c_{L1} & c_{L2} & c_{L3} & \cdots & c_{Ll} & \cdots & 0
\end{bmatrix}. \tag{6.1}
$$

Therefore, the overall load of each subarea belongs to the following set:

$$
\begin{aligned}
S &= \left\{ \sum_{l=1}^{L} c_{l1}, \sum_{l=1}^{L} c_{l2}, \sum_{l=1}^{L} c_{l3}, \ldots, \sum_{l=1}^{L} c_{lL} \right\} \\
&= \{ S_1, S_2, S_3, \ldots, S_l, \ldots S_L \}.
\end{aligned} \tag{6.2}
$$

Sort the elements in set S in a descending order, and then extract the first W elements from it, expressed as

$$
Max\_value = \{ M_1, M_2, M_3, \ldots, M_W \}. \tag{6.3}
$$

The first r elements in set Max_value are selected as the hot spot location (HSL) to deploy RSUs, we thus have

$$
HSL\,(r) = \{ 1, 2, 3, \ldots, r \}. \tag{6.4}
$$

In this way, the optimal locations to deploy RSUs can be determined.

6.2.2   Attack and Defence in Vehicular Networks

In vehicular networks, both vehicles and RSUs are vulnerable to various attacks from the outside or inside networks, violating the security of the applications in the networks [33–36]. The attacker can be an outsider independent of the network, or a person disguised as an ordinary vehicle user [37]. External attackers will first wait within the vehicle's wireless communication range, and when the time is ripe, they will hold one or more vehicles to make them compromise, so that they can have a legitimate identity for network destruction. Attackers can eavesdrop, block or discard communications between nodes within the network communication range. An attacker mainly aims to forbid normal data transmission, forging or modifying data, or deliberately submitting forged data to undermine network security. Specifically, literature [38] introduces three main attack modes in vehicular networks:

- Single Attack: An attacker can manipulate compromised nodes into violating the communication protocols and make them do not offer assistance to other nodes. For example, they will not respond to other nodes' requests in the networks. However, when asked about the trustworthiness of other nodes, the potential malicious node will give false advices about other nodes.
- Bad Mouth Attack: In addition to the single attacks described above, attackers will hide their identities and behave normally to deceive other nodes. In this way, the malicious actions of them will be hard to be detected.
- Zigzag Attack: In general, the cunning attackers can change their malicious attack mode to make it more difficult to be detected by trust management solutions. For example, they can conduct malicious attacks in a certain duration. In this case, the malicious attack is performed in a switch mode. In addition, some malicious attackers can also behave different to different nodes, which may lead to different advices of different witnesses on the same node's trust evaluation. Because there is not enough evidence to accuse malicious attackers, it is more difficult for the nodes in the networks to identify such cunning attackers.

During the interaction between the vehicle and the RSU, due to the open network of vehicular networks, both the vehicle and the RSU are vulnerable to be hijacked by attackers. When they interact with each other, they will maliciously spread false information. Therefore, to maintain network security, we need to add a protection mechanism. In vehicular networks, we assume that neither the vehicle user nor the RSU is completely trusted [39–43]. The communication channel is also vulnerable to be attacked due to the open structure of the network. Therefore, the adversary or network hijacker in vehicular networks can implement the following attacks without a trust management mechanism:

- In the process of V2V communication or V2I communication, a cyber hijacker may modify the above-mentioned information or directly replace the existing information with false information. The hijackers may also imitate the behavior of a legitimate vehicle user or RSU and send false information to other users in the networks. These behaviors will affect the behaviors of other users in the network, damage the infrastructure in the network, and threaten the communication security of the network.
- When the vehicle communicates with other vehicles or RSUs, the personal privacy information of the vehicle user, such as the current location, destination, hobbies, etc., may be involved in the information transmission process. Such information may be intercepted by the network hijacker during the transmission of the information. If the vehicle user or RSU is hijacked by a network hijacker, the personal privacy information of a large number of vehicle users may be leaked.

Therefore, in order to solve the problem of network security threats, in this section, we design the method of trust evaluation to evaluate the security behavior of vehicle users and RSUs in the networks. The vehicle users with low trust values or who used to threaten the networks will be banned to access the networks. Specific embodiments are shown below.

6.2.3 Trust Evaluation of Vehicles

To evaluate the trust of vehicles, we first add AU to the network, where the AU is a type of authoritative node. It can be composed of a series of vehicles with public service significance, such as police cars, buses, road rescue vehicles, and so on. There are several benefits by adding mobile AUs to the network:

- AU has higher authority than normal vehicles.
- AU can monitor the safety of vehicles and RSUs at the same time.
- Compared with the infrastructures fixed on the roadside, AU is more flexible, and it is easier to interact with more RSUs and vehicles.
- Since AU does not need to obtain benefits or services from the network like other users, it has no subjective motivation to attack the network and is not vulnerable to adversary attacks.

Then AU conducts the safety assessment of an information interaction between the vehicle and RSU. The vehicle and RSU forward the data information generated during the interaction to the AU. The AU then evaluates the information to judge the trust values of the vehicle and RSU.

The system model is shown in Fig. 6.3, where security events are recorded by AUs [44] when AUs interact with vehicles and RSUs. The sets of vehicles and AUs are denoted by $\{1, \ldots, v, \ldots, V\}$ and $\{1, \ldots, n, \ldots, N\}$. The trust value of the vehicle will be updated after interacting with AU. It is defined by

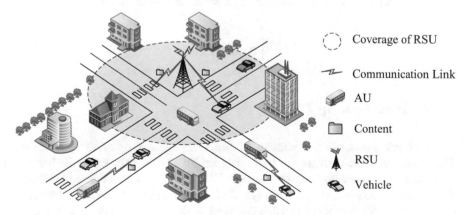

Coverage of RSU

Communication Link

AU

Content

RSU

Vehicle

Fig. 6.3 System model

$$SC_v = \left(1 - \frac{\sum_{t=0}^{t_{v,n}} n_{dd} + n_{neg}}{\Omega} \right) \times (1 - \Gamma_v), \tag{6.5}$$

where n_{dd} and n_{neg} denote the number of dangerous driving and negative in transmitting contents before this time interaction. The evaluation indicators are as follows.

- **Driving safety:** if the user has violated regulations in the driving process such as speeding, running a red light, etc., its trust value will be decreased. The more the number of violations, the greater the decline in the user's trust values.
- **Network participation:** when there are some emergency situations in the network, such as a car accident that causes the road to be blocked, vehicle A or an RSU finds this situation and forwards the traffic information to vehicle B. However, vehicle B does not forward this information to other nodes in the networks due to the fear of leaking its own geographic location or other personal privacy. This situation will make vehicle B be judged to be passively participating in network activities, resulting in the decline of its trust value.
- **Malicious behaviors:** if a vehicle user is hijacked by a malicious adversary, in order to intentionally undermine the security of the network, it will maliciously occupy the load of the RSU. In particular, it may send a large number of requests to the RSU within a certain period of time. After hijacking the RSU, a lot of false information is sent to other vehicle nodes and RSUs. Once the vehicle user with these malicious behaviors is detected by the AU, the trust value of the vehicle will be directly cleared and it will be banned to access the network.

The count of these two numbers should be less than Ω, or the trust value of v would be 0. Let A=1 denote the user who has malicious actions (e.g., disseminate malicious contents). Γ_v is the judgment of such actions, it can be expressed as

$$\Gamma_v = \begin{cases} 1, & A = 1, \\ 0, & otherwise. \end{cases} \tag{6.6}$$

6.2.4   Trust Value of RSUs

The trust assessment of RSUs consists of direct and indirect trust value assessment. The direct trust value assessment is the trust evaluation by AUs, while the indirect trust evaluation is accomplished by vehicles. The direct trust evaluation of RSUs is defined by the proportion of the number of contents that the RSUs safely deliver to the vehicles to the number of requests received by the RSUs from the vehicles. If an RSU has malicious actions like transmitting a large number of low quality contents to vehicles, its direct trust value will be set to 0.

RSUs can be shown by $\{1, \ldots, r, \ldots, R\}$. For RSU r, its direct trust value will be updated after interacting with AU n. It is given by

$$DSC_r = \frac{\sum\limits_{t=0}^{t_{r,n}} \sum\limits_{v=1}^{V} EVA_r(v)}{\sum\limits_{t=0}^{t_{r,n}} \sum\limits_{v=1}^{V} Q_r(v)} \times (1 - \Gamma_r), \tag{6.7}$$

where Γ_r is the judgment value of the malicious actions. $\sum\limits_{t=0}^{t_{r,n}} \sum\limits_{v=1}^{V} EVA_r(v)$ and $\sum\limits_{t=0}^{t_{r,n}} \sum\limits_{v=1}^{V} Q_r(v)$ denote the evaluation numbers and the query numbers that r has received from vehicles before this time interaction.

The stay time of v stays in r's range is T_v. The opinion about the quality of content given by v is $e_{r,v}(e_{r,v} \in E)$, where $E = \{-2, -1, 0, 1, 2\}$. $IDSC_r^b$ is the indirect trust value of r before giving this opinion, we have

(a) If $e_{r,v} > 0$, the indirect trust value increases

$$\Delta IDSC_{r,v} = \zeta e_{r,v} IDSC_r^b. \tag{6.8}$$

ζ is adjustment coefficient.

(b) If $e_{r,v} < 0$, the indirect trust value decreases

$$\Delta IDSC_{r,v} = -\zeta s_r e_{r,v} IDSC_r^b. \tag{6.9}$$

Here, s_r is the numbers that gets minus the evaluation value.

(c) If $e_{r,v}$ equals to 0, the indirect trust value will not be increased or decreased.

According to (6.8) and (6.9), the indirect trust value of r is

$$IDSC_r = IDSC_r^b + \Delta IDSC_{r,v} \tag{6.10}$$

$$= \begin{cases} (1 + \zeta e_{r,v})IDSC_r^b, & e_{r,v} > 0, \\ (1 - \zeta s_r e_{r,v})IDSC_r^b, & e_{r,v} < 0, \\ IDSC_r^b, & e_{r,v} = 0, \end{cases}$$

where $0 \leq IDSC_r^b, IDSC_r \leq 1$.

According to (6.7) and (6.10), the trust value of r is

$$SC_r = \frac{DSC_r + IDSC_r}{2} \tag{6.11}$$

$$= \frac{\sum\limits_{t=0}^{t_{r,n}} \sum\limits_{v=1}^{V} EVA_r(v)}{2 \sum\limits_{t=0}^{t_{r,n}} \sum\limits_{v=1}^{V} Q_r(v)} \times (1 - \Gamma_r) + \frac{1}{2} \begin{cases} (1 + \zeta e_{r,v})IDSC_r^b, & e_{r,v} > 0, \\ (1 - \zeta s_r e_{r,v})IDSC_r^b, & e_{r,v} < 0, \\ IDSC_r^b, & e_{r,v} = 0. \end{cases}$$

6.2.5   Deployment of AU

Because the load conditions of different areas in vehicular networks mentioned above are different, in order to quickly and fully evaluate the trust values of nodes in the network, we need to specifically consider the number of AUs deployed in each area. The AU is composed of a series of vehicles with public service significance, such as police cars, buses, and road rescue vehicles. Considering that buses have a fixed driving route and road rescue vehicles will only appear at certain special times, we only consider the deployment of a class of police vehicles to evaluate the trust values of network nodes.

We consider each RSU's communication range as an independent area. The traffic load in these r areas belongs to the set denoted as

$$S_r = \{M_1, M_2, M_3, \ldots, M_r\}. \tag{6.12}$$

In this way, the ratio of the traffic load in the m-th ($m = \{1, 2, \ldots, r\}$) area to the total traffic load in r areas can be denoted by

$$\eta_m = \frac{M_i}{\sum\limits_{j=1}^{r} M_j}. \tag{6.13}$$

According to (6.13), the amount of AUs deployed in area m is positively correlated to η_m.

Although AU is a type of authoritative node and does not have a subjective motive for attacking the network, AU still has a certain probability of being hijacked by adversaries. The hijacking of AU may occur deliberately sending a large number of low ratings to a user to damage the network safety. In view of this situation, we can limit the maximum number of times that an AU evaluates the same user and then introduces an appeal mechanism. Specifically, if the vehicle user thinks that the AU is malicious who will deliberately give a low evaluation value, it can appeal to the higher-level supervisory authority, and another AU will perform a secondary evaluation of the user.

6.2.6   Bargain Game Between RSUs and Vehicles

To incent vehicles and RSUs to enhance their trustworthiness and obtain more utilities, in this subsection, we present a bargain game between them [45–49]. Firstly, we give the utility function of RSUs by the lowest reservation price, shown as

$$P_r = P\left[1 + \alpha L_r + \beta(SC_r - 0.5)\right]$$

$$= P\left[1 + \alpha \sum_{v=1}^{V} L_r(v)\right]$$

$$+ \beta P\left[\frac{\sum\limits_{t=0}^{t_{r,n}}\sum\limits_{v=1}^{V} EVA_r(v)}{2\sum\limits_{t=0}^{t_{r,n}}\sum\limits_{v=1}^{V} Q_r(v)} \times (1 - \Gamma_v) - 0.5\right] + \frac{\beta P}{2}\left\{\begin{array}{l} (1 + \zeta e_{r,v})IDSC_r^b, \ e_{r,v} > 0. \\ (1 - \zeta s_r e_{r,v})IDSC_r^b, , \ e_{r,v} < 0. \\ IDSC_r^b, \ e_{r,v} = 0, \end{array}\right.$$

$$(6.14)$$

where P is the basic price, α and β are adjustment coefficient. If r's trust value is lower than 0.5, the price of it will be decreased with the target of obtaining a higher trust value. L_r is the load of r. It can be defined as

$$L_r = \sum_{v=1}^{V} L_r(v). \qquad (6.15)$$

Let $B = 1$ denote that v is connecting with r, we have

$$L_r(v) = \begin{cases} 1, \ B = 1, \\ 0, \ otherwise. \end{cases} \qquad (6.16)$$

Next, we present vehicle's utility function. A vehicle should select an appropriate RSU to obtain content in time. In other words, a vehicle may not obtain content before it leaves the RSU's range due to high load of that RSU. To relieve the pressure of RSU with high load, we propose the threshold of RSU's load. A vehicle which intends to connect to an RSU can select the one with a lower load than the threshold value.

We suppose the velocity of vehicle v is a constant value, denoted as f_v. Then, the stay time of v in r's range is

$$T_v = \frac{D}{f_v}, \qquad (6.17)$$

where D is the diameter of r's range.

The transmission time of content between r and v can be denoted by

$$t_{r,v} = t_0 (1 + L_r).$$ (6.18)

Here, t_0 is the average time that r transfers information to a vehicle.

According to (6.17) and (6.18), the load limit of r is shown as

$$L_r \leq \frac{D}{f_v t_0} - 1.$$ (6.19)

Then, we present vehicle's utility function by giving the highest reservation price, which is denoted by

$$P_v = P_1 \left[1 + \omega \left(T_v - t_{r,v} \right) + \lambda \left(SC_r - 0.5 \right) + \rho \left(0.5 - SC_v \right) \right],$$ (6.20)

where P_1 is the basic price for all vehicles. ρ, ω and λ are adjustment factors.

According to (6.5), (6.11) and (6.15), (6.20) can be rewritten as:

$$
\begin{aligned}
P_v = P_1 \Bigg\{ 1 + \rho \Bigg[0.5 &- \left(1 - \frac{\sum_{t=0}^{t_{v,n}} n_{dd} + n_{neg}}{10} \right) \times (1 - \Gamma_v) \Bigg] \\
&+ \omega \left[T_v - t_0 \left(1 + \sum_{v=1}^{V} L_r(v) \right) \right] \Bigg\} \\
&+ \lambda P_1 \left[\frac{\sum_{t=0}^{t_{r,n}} \sum_{v=1}^{V} EVA_r(v)}{2 \sum_{t=0}^{t_{r,n}} \sum_{v=1}^{V} Q_r(v)} \times (1 - \Gamma_v) - 0.5 \right] \\
&+ \frac{\lambda P_1}{2} \begin{cases} (1 + \zeta e_{r,v}) IDSC_r^b, & e_{r,v} > 0, \\ (1 - \zeta s_r e_{r,v}) IDSC_r^b, & e_{r,v} < 0, \\ IDSC_r^b, & e_{r,v} = 0. \end{cases}
\end{aligned}
$$ (6.21)

If $P_v > P_r$, the bargain game will begin between v and r. In this game, the seller S (i.e., RSU r) offers a price between P_r and P_v first. If the buyer B (i.e., vehicle v) doesn't accept this price, it will offer another price between P_r and P_v. The utility functions of r and v can be expressed as

$$u_r(x_r) = x_r C,$$ (6.22)

$$u_v(x_v) = x_v C,$$ (6.23)

where $C = P_v - P_r$ denotes the differential of these two prices and $x_r + x_v = 1$.

With the game's persistence, the benefits of both r and v have a discount in each round, denoted by δ $(0 < \delta < 1)$. The discount coefficient of r and v, i.e., δ_r and δ_v can be shown as

$$\delta_v = \frac{e^{v\theta} - e^{-v\theta}}{e^{v\theta} + e^{-v\theta}}, \tag{6.24}$$

$$\delta_r = \frac{e^{\mu\theta_1} - e^{-\mu\theta_1}}{e^{\mu\theta_1} + e^{-\mu\theta_1}}, \tag{6.25}$$

where θ presents the stay time of v in r's coverage minus the waiting time of the contents. θ_1 is load of r. We have

$$\begin{cases} \theta = T_v - t_{r,v} = T_v - t_0\,(1 + L_r), \\ \frac{d\delta_v(\theta)}{d\theta} > 0, \end{cases} \tag{6.26}$$

$$\begin{cases} \theta_1 = L_r = \sum\limits_{v=1}^{V} L_r\,(v), \\ \frac{d\delta_r(\theta_1)}{d\theta_1} > 0. \end{cases} \tag{6.27}$$

We then analyze the situation that the game is not finished in the first round. The utility functions of the two participants in k-th round are

$$u_r^k\left(x_r^k\right) = \delta_r^{k-1} x_r^k C, \tag{6.28}$$

$$u_v^k\left(x_v^k\right) = \delta_v^{k-1} x_v^k C. \tag{6.29}$$

The infinite rounds game can be changed by a three-round game. We suppose x^* is the optimal allocation scheme belongs to RSU r. As such, $1 - x^*$ is the share that r can be earned from C in the third round. We have

$$\begin{cases} x_r^3 = x^*, \\ x_v^3 = 1 - x^*. \end{cases} \tag{6.30}$$

Therefore, the utilities of r and v in the third round can be derived as

$$\begin{cases} u_r^3 = \delta_r^2 x^* C, \\ u_v^3 = \delta_v^2\,(1 - x^*)\,C. \end{cases} \tag{6.31}$$

Then we move to the second round, where the allocation mechanism is made by v. We have

$$\begin{cases} x_r^2 = x, \\ x_v^2 = 1 - x. \end{cases} \tag{6.32}$$

The utilities of r and v in the second round can be given by

$$\begin{cases} u_r^2 = \delta_r x C, \\ u_v^2 = \delta_v (1 - x) C. \end{cases} \tag{6.33}$$

If $u_r^2 \geq u_r^3$, i.e., $x \geq \delta_r x^*$, r accepts the allocation scheme that is proposed by v in the second round and the game ends in this round. Otherwise, the game will continue to the next round. The optimal allocation scheme of v in this round can be denoted by $x = \delta_r x^*$.

Next, we go back to the first round, where the allocation scheme is made by r. We have

$$\begin{cases} x_r^1 = x^*, \\ x_v^1 = 1 - x^*. \end{cases} \tag{6.34}$$

The utilities of r and v in the first round can be expressed as

$$\begin{cases} u_r^1 = x^* C, \\ u_v^1 = (1 - x^*) C. \end{cases} \tag{6.35}$$

When $u_v^1 \geq u_v^2$, i.e. $1 - x^* \geq \delta_v (1 - x)$, v accepts the allocation scheme that r proposed in the first round. The game is ended in this round, which leads the optimal utilities to both r and v. The optimal allocation scheme can be denoted by $1 - x^* = \delta_v (1 - x)$.

Thus we can get the subgame perfect Nash equilibrium of the game by

$$x_r^* = \frac{1 - \delta_v}{1 - \delta_v \delta_r}, \tag{6.36}$$

$$x_v^* = \frac{\delta_v - \delta_v \delta_r}{1 - \delta_v \delta_r}. \tag{6.37}$$

The transaction price of the bargain game can be denoted by

$$P^* = P_r + u_r \left(x_r^* \right) \tag{6.38}$$

$$= P_r + \frac{1 - \delta_v}{1 - \delta_v \delta_r} C$$

$$= \frac{\delta_v - \delta_v \delta_r}{1 - \delta_v \delta_r} P_r + \frac{1 - \delta_v}{1 - \delta_v \delta_r} P_v.$$

By substituting P_r and P_v in (6.14) and (6.20) into (6.38), P^* is shown as

$$
P^* = \frac{\delta_v - \delta_v \delta_r}{1 - \delta_v \delta_r} P \left(1 + \alpha \sum_{v=1}^{V} L_r(v) \right) + \frac{1 - \delta_v}{1 - \delta_v \delta_r}
$$

$$
\cdot P_1 \left\{ 1 + \omega \left[T_v - t_0 \left(1 + \sum_{v=1}^{V} L_r(v) \right) \right] \right\}
$$

$$
- \frac{1 - \delta_v}{1 - \delta_v \delta_r} \rho P_1 \left[0.5 - \left(1 - \frac{\sum_{t=0}^{t_{v,n}} n_{dd} + n_{neg}}{10} \right) \times (1 - \Gamma_v) \right]
$$

$$
+ \left(\frac{\delta_v - \delta_v \delta_r}{1 - \delta_v \delta_r} \beta P + \frac{1 - \delta_v}{1 - \delta_v \delta_r} \lambda P_1 \right)
$$

$$
\cdot \left[\frac{\sum_{t=0}^{t_{r,n}} \sum_{v=1}^{V} EVA_r(v)}{2 \sum_{t=0}^{t_{r,n}} \sum_{v=1}^{V} Q_r(v)} \times (1 - \Gamma_r) - 0.5 \right]
$$

$$
+ \left(\frac{\delta_v - \delta_v \delta_r}{1 - \delta_v \delta_r} \frac{\beta P}{2} + \frac{1 - \delta_v}{1 - \delta_v \delta_r} \frac{\lambda P_1}{2} \right) \cdot \begin{cases} (1 + \zeta e_{r,v}) IDSC_r^b, & e_{r,v} > 0, \\ (1 - \zeta s_r e_{r,v}) IDSC_r^b, & e_{r,v} < 0, \\ IDSC_r^b, & e_{r,v} = 0. \end{cases}
$$

$$(6.39)$$

6.3 Simulation

6.3.1 Setting

In the simulation, we evaluate the effect of the RSU's load on the RSU's and vehicle's utilities. The parameters used in the simulation are summarized in Table 6.1. We compare the performance of our proposal with a conventional scheme called Random Offer Price (ROP). In the ROP scheme, the price of content delivery service are randomly offered by vehicles and RSUs.

6.3.2 Results Analysis

Figure 6.4 shows the results of the vehicle's and the RSU's utilities in the aforementioned schemes by changing the load of the RSU from 0 to 9. From Fig. 6.4, we can see that both the utilities of the vehicle and the RSU are decreased

Table 6.1 Simulation parameters

Parameters	Values
P, P_1	10, 10
α, ω	1, 1
SC_r, SC_v	0.5, 0.5
T_r	20
t_0	1
μ, υ	0.04, 0.04
P_r, P_v	[5, 115],[5, 115]

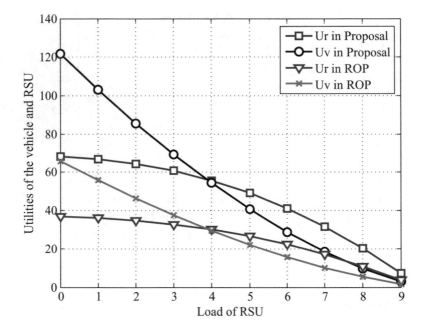

Fig. 6.4 Utilities of the vehicle and the RSU by changing the load of RSU

with the increase of the RSU's load. The reasons are firstly, the price offered by vehicle decreases when the RSU's load becomes larger which may lead to more waiting time. Secondly, the price offered by the RSU increases accordingly if the load of the RSU is high. Moreover, compared with the ROP scheme, both the vehicle and the RSU can obtain higher utilities in the proposed scheme.

6.4 Summary

In this chapter, we have presented a secure content delivery mechanism to protect nodes from various threats in vehicular networks. In the proposed scheme, we have introduced a novel method to assess trust values of vehicles and RSUs by introducing AUs into the networks. Then, to obtain higher utilities for the nodes and enhance their trust values, we have presented a competitive scheme with the adoption of bargain game to improve their utilities. The simulation result has shown that the proposed scheme has a higher efficiency than the conventional method.

References

1. C. Bila, F. Sivrikaya, M.A. Khan, S. Albayrak, Vehicles of the future: a survey of research on safety issues. IEEE Trans. Intell. Transp. Syst, **18**(5), 1046–1065 (2017)
2. Z. Su, Y. Hui, Q. Yang, The next generation vehicular networks: a content-centric framework. IEEE Wirel. Commun. **24**(1), 60–66 (2017)
3. T.H. Luan, L.X. Cai, J. Chen, X.S. Shen, F. Bai, Engineering a distributed infrastructure for large-scale cost-effective content dissemination over urban vehicular networks. IEEE Trans. Veh. Technol. **63**(3), 1419–1435 (2014)
4. N. Cheng, F. Lyu, J. Chen, W. Xu, H. Zhou, S. Zhang, X. Shen, Big data driven vehicular networks. IEEE Netw. **32**(6), 160–167 (2018)
5. Y. Hui, Z. Su, T.H. Luan, Collaborative content delivery in software-defined heterogeneous vehicular networks. IEEE/ACM Trans. Netw. **28**(2), 575–587 (2020)
6. A. Dua, N. Kumar, S. Bawa, A systematic review on routing protocols for vehicular ad hoc networks. Veh. Commun. **1**(1), 33–52 (2014)
7. S. Zhang, J. Chen, F. Lyu, N. Cheng, W. Shi, X. Shen, Vehicular communication networks in the automated driving era. IEEE Commun. Mag. **56**(9), 26–32 (2018)
8. Y. Hui, Z. Su, T.H. Luan, J. Cai, A game theoretic scheme for optimal access control in heterogeneous vehicular networks. IEEE Trans. Intell. Transp. Syst. **20**(12), 4590–4603 (2019)
9. N. Cheng, N. Lu, N. Zhang, X. Zhang, X.S. Shen, J.W. Mark, Opportunistic WiFi offloading in vehicular environment: a game-theory approach. IEEE Trans. Intell. Transp. Syst. **17**(7), 1944–1955 (2016)
10. A. Bazzi, A. Zanella, G. Cecchini, B.M. Masini, Analytical investigation of two benchmark resource allocation algorithms for LTE-V2V. IEEE Trans. Veh. Technol. **68**(6), 5904–5916 (2019)
11. S.A. Ahmad, A. Hajisami, H. Krishnan, F. Ahmed-Zaid, E. Moradi-Pari, V2V system congestion control validation and performance. IEEE Trans. Veh. Technol. **68**(3), 2102–2110 (2019)
12. Y. Park, T. Kim, D. Hong, Resource size control for reliability improvement in cellular-based V2V communication. IEEE Trans. Veh. Technol. **68**(1), 379–392 (2019)
13. R. Atallah, M. Khabbaz, C. Assi, Multihop V2I communications: a feasibility study, modeling, and performance analysis. IEEE Trans. Veh. Technol. **66**(3), 2801–2810 (2017)
14. Y. Xu, D. Li, Y. Xi, A game-based adaptive traffic signal control policy using the vehicle to infrastructure (V2I). IEEE Trans. Veh. Technol. **68**(10), 9425–9437 (2019)
15. Y. Hui, Z. Su, T.H. Luan, J. Cai, Content in motion: an edge computing based relay scheme for content dissemination in urban vehicular networks. IEEE Trans. Intell. Transp. Syst. **20**(8), 3115–3128 (2019)
16. M. Yusuf, E. Tanghe, F. Challita, P. Laly, D.P. Gaillot, M. Liénard, L. Martens, W. Joseph, Stationarity analysis of V2I radio channel in a suburban environment. IEEE Trans. Veh. Technol. **68**(12), 11532–11542 (2019)
17. G. Li, L. Boukhatem, J. Wu, Adaptive quality-of-service-based routing for vehicular ad hoc networks with ant colony optimization. IEEE Trans. Veh. Technol. **66**(4), 3249–3264 (2017)
18. Z. Su, Y. Hui, Q. Xu, T. Yang, J. Liu, Y. Jia, An edge caching scheme to distribute content in vehicular networks. IEEE Trans. Veh. Technol. **67**(6), 5346–5356 (2018)

19. S. Abdelhamid, H.S. Hassanein, G. Takahara, On-road caching assistance for ubiquitous vehicle-based information services. IEEE Trans. Veh. Technol. **64**(12), 5477–5492 (2015)
20. H. Zhou, N. Cheng, J. Wang, J. Chen, Q. Yu, X. Shen, Toward dynamic link utilization for efficient vehicular edge content distribution. IEEE Trans. Veh. Technol. **68**(9), 8301–8313 (2019)
21. Z. Su, Y. Hui, S. Guo, D2D based content delivery with parked vehicles in vehicular social networks. IEEE Wirel. Commun. **23**(4), 90–95 (2016)
22. Y. Zhou, F.R. Yu, J. Chen, Y. Kuo, Resource allocation for information-centric virtualized heterogeneous networks with in-network caching and mobile edge computing. IEEE Trans. Veh. Technol. **66**(12), 11339–11351 (2017)
23. J. Kang, R. Yu, X. Huang, M. Wu, S. Maharjan, S. Xie, Y. Zhang, Blockchain for secure and efficient data sharing in vehicular edge computing and networks. IEEE Internet Things J. **6**(3), 4660–4670 (2019)
24. K. Fan, X. Wang, K. Suto, H. Li, Y. Yang, Secure and efficient privacy-preserving ciphertext retrieval in connected vehicular cloud computing. IEEE Netw. **32**(3), 52–57 (2018)
25. H. Guo, Z. Zhang, J. Liu, FiWi-enhanced vehicular edge computing networks: collaborative task offloading. IEEE Veh. Technol. Mag. **14**(1), 45–53 (2019)
26. C. Lin, D. Deng, C. Yao, Resource allocation in vehicular cloud computing systems with heterogeneous vehicles and roadside units. IEEE Internet Things J. **5**(5), 3692–3700 (2018)
27. N. Zhang, N. Cheng, A.T. Gamage, K. Zhang, J.W. Mark, X. Shen, Cloud assisted HetNets toward 5G wireless networks. IEEE Commun. Mag. **53**(6), 59–65 (2015)
28. C. Lai, H. Zhou, N. Cheng, X.S. Shen, Secure group communications in vehicular networks: a software-defined network-enabled architecture and solution. IEEE Veh. Technol. Mag. **12**(4), 40–49 (2017)
29. A. Yang, J. Weng, N. Cheng, J. Ni, X. Lin, X. Shen, Deqos attack: degrading quality of service in vanets and its mitigation. IEEE Trans. Veh. Technol. **68**(5), 4834–4845 (2019)
30. C. Lai, K. Zhang, N. Cheng, H. Li, X. Shen, SIRC: A secure incentive scheme for reliable cooperative downloading in highway VANETs. IEEE Trans. Intell. Transp. Syst. **18**(6), 1559–1574 (2017)
31. C.A. Kerrache, C.T. Calafate, J.C. Cano, N. Lagraa, P. Manzoni, Trust management for vehicular networks: an adversary-oriented overview. IEEE Acc. **4**, 9293–9307 (2016)
32. A. Wasef, R. Lu, X. Lin, X. Shen, Complementing public key infrastructure to secure vehicular ad hoc networks: security and privacy in emerging wireless networks. IEEE Wirel. Commun. **17**(5), 22–28 (2010)
33. J.A. Onieva, R. Rios, R. Roman, J. Lopez, Edge-assisted vehicular networks security. IEEE Internet Things J. **6**(5), 8038–8045 (2019)
34. Y. Yao, X. Chang, J. Misic, V. Misic, Reliable and secure vehicular fog service provision. IEEE Internet Things J. **6**(1), 734–743 (2019)
35. F. Tang, Y. Kawamoto, N. Kato, J. Liu, Future intelligent and secure vehicular network toward 6G: machine-learning approaches. Proc. IEEE **108**(2), 292–307 (2020)
36. D. Wang, P. Ren, Q. Du, L. Sun, Y. Wang, Security provisioning for miso vehicular relay networks via cooperative jamming and signal superposition. IEEE Trans. Veh. Technol. **66**(12), 10732–10747 (2017)
37. J. Feng, N. Liu, J. Cao, Y. Zhang, G. Lu, Securing traffic-related messages exchange against inside-and-outside collusive attack in vehicular networks. IEEE Internet Things J. **6**(6), 9979–9992 (2019)
38. W. Li, H. Song, Art: an attack-resistant trust management scheme for securing vehicular ad hoc networks. IEEE Trans. Intell. Transp. Syst. **17**(4), 960–969 (2016)
39. H. Xia, S. Zhang, Y. Li, Z. Pan, X. Peng, X. Cheng, An attack-resistant trust inference model for securing routing in vehicular ad hoc networks. IEEE Trans. Veh. Technol. **68**(7), 7108–7120 (2019)
40. Z. Yang, K. Yang, L. Lei, K. Zheng, V.C.M. Leung, "Blockchain-based decentralized trust management in vehicular networks. IEEE Internet Things J. **6**(2), 1495–1505 (2019)

41. H. Hu, R. Lu, Z. Zhang, J. Shao, Replace: a reliable trust-based platoon service recommendation scheme in vanet. IEEE Trans. Veh. Technol. **66**(2), 1786–1797 (2017)
42. G. Han, Y. He, J. Jiang, N. Wang, M. Guizani, J.A. Ansere, A synergetic trust model based on svm in underwater acoustic sensor networks. IEEE Trans. Veh. Technol. **68**(11), 11239–11247 (2019)
43. B. Luo, X. Li, J. Weng, J. Guo, J. Ma, Blockchain enabled trust-based location privacy protection scheme in vanet. IEEE Trans. Veh. Technol. **69**(2), 2034–2048 (2020)
44. C. Chen, W.L. Han, L. Zhu, X. Wang, Research for reputation models in vanet. J. Chin. Comput. Syst. **34**(2), 233–237 (2013)
45. H. Zhang, C. Jiang, N.C. Beaulieu, X. Chu, X. Wang, T.Q.S. Quek, Resource allocation for cognitive small cell networks: a cooperative bargaining game theoretic approach. IEEE Trans. Wirel. Commun. **14**(6), 3481–3493 (2015)
46. L. Gao, G. Iosifidis, J. Huang, L. Tassiulas, D. Li, Bargaining-based mobile data offloading. IEEE J. Sel. Areas Commun. **32**(6), 1114–1125 (2014)
47. Y. Hui, Z. Su, S. Guo, Utility based data computing scheme to provide sensing service in internet of things. IEEE Trans. Emerg. Top. Comput. **7**(2), 337–348 (2019)
48. H. Yuan, X. Wei, F. Yang, J. Xiao, S. Kwong, Cooperative bargaining game-based multiuser bandwidth allocation for dynamic adaptive streaming over HTTP. IEEE Trans. Multimedia **20**(1), 183–197 (2018)
49. X. Cao, J. Zhang, H.V. Poor, Data center demand response with on-site renewable generation: a bargaining approach. IEEE/ACM Trans. Netw. **26**(6), 2707–2720 (2018)

Chapter 7
Deep Learning Based Autonomous Driving in Vehicular Networks

When kids, we may have dreamed of playing video games and watching movies on the trip with our parents, while letting the cars drive by themselves. This finally becomes practical with the autonomous driving, which relies on the vehicles to sense and learn the environment and determine the driving behavior without or with few human operations. This, however, is quite challenging due to the huge data perceived from complicated traffic environment to be analyzed in real-time and the limited computing power of vehicles. In this chapter, we provide a brief survey on the state-of-the-art autonomous driving technology. Towards this goal, we present the basic structure of hardware and software modules of autonomous vehicles and the application of deep learning in autonomous driving. In particular, note that by using wireless communications to connect vehicles as a network, the autonomous vehicles can share the information and collaboratively adjust the driving behaviors. We further propose a collaborative driving framework in which autonomous vehicles learn and drive with groups. Using simulations, we show how wireless communications can help with collaborative sensing and deep learning in autonomous driving.

7.1 Introduction

A few years ago, Google launched its self-driving project Waymo and brought autonomous driving into the sight of the public. Nowadays, the autonomous driving has already been enthusiastically pursued worldwide by most car manufacturers and governments [1–5]. For example, some car manufacturers (e.g., Tesla, Toyota and Audi) have all investigated on artificial intelligence and released their self-driving models. The IT companies (e.g., Google, Nvidia and Baidu) have also released autonomous driving operating systems. The US has established ten official test fields for autonomous driving and the Obama Administration has announced that

© Springer Nature Switzerland AG 2021

Z. Su et al., *The Next Generation Vehicular Networks, Modeling, Algorithm, and Applications*, Wireless Networks, https://doi.org/10.1007/978-3-030-56827-6_7

it would request a $4 billion investment over 10 years for technology research and infrastructure improvements related to autonomous vehicles (AVs) [6].

The autonomous driving attracts the eyes of the whole world for the following reasons:

- **Reduce death:** A report provided by the Association for Safe International Road Travel shows that nearly 1.3 million people die in road crashes each year, where the main reason that causes accidents is attributable to the inappropriate operations of drivers [7]. The autonomous driving can significantly reduce accidents and death as the self-driving cars will not be distracted like a human being.
- **Improve efficiency:** The rapid increase in vehicles has worsened the problems of traffic and parking. These, not only waste people's time but also result in high energy consumption and pollution. With autonomous driving, AVs can drive more efficiently than before, where the traffic congestion can be decreased and the parking problem can be alleviated.
- **Enhance experience:** The autonomous driving can not only free drivers from getting into the boring driving activities, but also can provide opportunities to people who are unable to drive, such as children, the aged or people with disabilities. As a result, drivers can obtain the enhanced experience by reading, resting, working even playing games in their cars.

While the autonomous driving is coming and its practice has already been proved in many field tests, it still faces fundamental challenges and needs significant improvements to be completely accepted by public.

- **Hardware:** The sensors equipped in AVs are limited in many aspects, e.g., perception range, accuracy and response time. Using the sensors alone may lead to errors and severe accidents, especially in complex traffic scenarios. For example, in 2016, a Tesla Model S was using the autopilot function to drive on the road in Florida. Because the collected data was incomplete, the AV could not identify the white truck ahead of it and crashed into the truck directly.
- **Data:** The data perceived by various sensors as well as the rich image information collected by the high-definition video camera are huge, complicated and noisy. For example, if an AV uses eight cameras which usually work at the frequency of 60 Hz to identify and track objects from around four directions, it will generate huge amounts of data (about 1.8 GB data per second). When an AV meets the bad weather, where there are suspended particles (e.g., raindrops or dust) in the air, the measurement of lidar will be greatly interfered with.
- **Software:** The data infused into the central processor unit (CPU) in AVs are irregular, unstructured and of high uncertainty and diversity. It is therefore difficult to extract feature information effectively for modeling the autonomous driving system. Even if a few feature parameters can be designed based on the prior knowledge, it is difficult to achieve the desired decision accuracy.

The wireless vehicular networks [8–14] and deep learning are two pillar technologies to address the above challenges towards efficient autonomous driving

[15–19]. In specific, the deep learning technology inputs the sensed environment data into the trained model and generates the output, which is then used by the CPU of autonomous driving to make driving decisions. In this way, deep learning automatically learns features of the road environment from a set of time-varying sensed data rather than designing manually, which can contain a large number of parameters to seek the potential rules in the data and make accurate decisions for driving. The wireless vehicular networks allow vehicles to timely communicate with the neighboring vehicles or infrastructures, and accordingly extend the hearing area of an AV. In addition, the AVs can collect road information from other vehicles to speed up their training process in the deep learning process. As a result, the AVs can acquire a much better understanding of the road environment in a much faster and cost-effective way, and make behavior decisions more efficiently and accurately than acting alone.

In this chapter, we discuss on how wireless technologies and deep learning can be combined to make autonomous driving. To this end, we first introduce the structure of an AV and surveys the state of the art technologies in autonomous driving. After that, we describe a collaborative autonomous driving framework which applies the wireless vehicular networks in the deep learning process. Lastly, we demonstrate the framework using extensive simulations.

7.2 Overview of Deep Learning Based Autonomous Driving in Vehicular Networks

7.2.1 Autonomous Driving

Autonomous driving (Full automated) is a process that the AV drives from the source to the destination without human participation. In this way, a driver only needs to provide the destination that he/she intends to go and the AV can drive to the destination automatically [20–24]. Apparently, it can make our trip more convenient and improve the quality of our life. According to the 2016 SAE International standard J3016, the levels of automation of AVs range from 0 to 5 as shown in Fig. 7.1 [25]. Although Baidu has launched the autonomous driving service at L4 in Wuzhen scenic spot in China and Audi also claims to launch L4 AVs in 2021, it's still a long way to achieve fully autonomous driving at L5. Currently, as shown in Fig. 7.2, an AV consists of the following basic components:

- **Sensors and cameras**: (1) Lidar: The lidar, which is usually installed on the roof of an AV, is used to obtain the point cloud data of the surrounding environment through high-speed rotation. It can identify the outline of objects and draw the three-dimensional spatial map around the AV in real time. In addition, the lidar also can measure the speed, acceleration and angular velocity of the nearby AVs. For example, Google uses the Velodyne's production HDL-64 which has a range of 120 m with 64 laser beams in its AV. Although lidar (e.g., HDL-64) has a

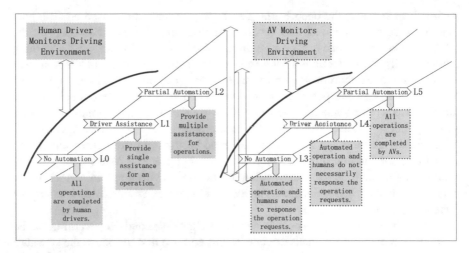

Fig. 7.1 The levels of autonomous driving

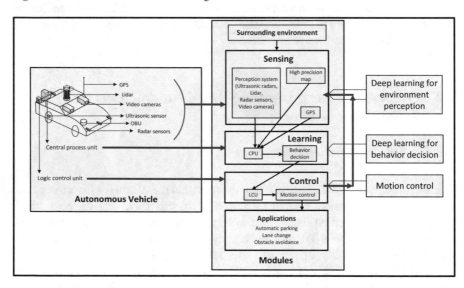

Fig. 7.2 The autonomous driving process

high resolution, high precision and strong anti-interference ability, its price is very expensive (about 80,000 dollars) so that each AV usually equipped with one lidar. (2) Ultrasonic radars: The ultrasonic radar uses the ultrasonic generator to produce the ultrasonic and then receives the ultrasonic wave reflected by the objects. According to the time difference between the delivered and received ultrasonics, the distance between the objects and the AV can be calculated. Although the ultrasonic radar has far transmission distance and the low cost, it has some limitations in measuring distance. First, the speed of ultrasonic transmission is easily affected by weather conditions. Second, because of the poor direction of ultrasonic waves, the precision is low in measuring the object of far distance. As a result, ultrasonic radar typically is used to measure the position of objects very close to the AV. (3) Radar sensors: The radar sensors can quickly and accurately obtain the environment information around an AV (e.g., the relative distance, relative velocity, angle and direction of motion) by transmitting the millimeter wave (between 1 and 10 mm) and receiving the target reflection signal. The range of the radar is not limited by weather conditions. But it has a small range of visibility with the result that each AV may have multiple radar sensors. (4) Video cameras: Camera is mainly used to collect images to estimate relative distance and relative speed of objects based on the moving mode of objects or binocular location. Cameras can also identify traffic signs and signals on the road to ensure that an AV's operation is strictly observed. However, the weather factors have a great influence on camera recognition. In addition, the camera has no penetration ability and can not detect the object being blocked.

- **GPS and high precision map**: Global position system (GPS) [26–31] is used to realize the self positioning of an AV. The positioning accuracy is low when purely relying on the GPS as the multi-path problem of GPS signal will cause noise interference. In addition, GPS must be used in a non-closed environment, so it can not be used in many scenarios, such as tunnels. High precision maps can provide the priori information to assist the self positioning and decision making. For example, the width of the road segment, the road sign, the course, the curvature of the road, the position of the speed limit mark, etc., can be obtained directly from the map. However, due to the high cost of data collection, a high precision map is usually difficult to obtain. In addition, even if the map is obtained, it is difficult to update it in real time to cope with unexpected changes.
- **CPU and LCU**: The CPU is mainly used to integrate and analyze the AV's own data and the surrounding environment information to make decision commands. The logic control unit (LCU) is mainly used to execute the control operation, namely, output the control behaviors to the brake, the throttle, the steering wheel of the AV etc., after receiving the control commands published by the CPU to achieve autonomous driving.

With these components, the autonomous driving process can be divided into the following steps.

- **Environment perception**: The AV uses a set of sensors and video cameras to sense, recognize and classify the surrounding environment including road signs, lanes, speed bumps, etc.
- **Behavior decision**: Behavior decision is the process that the CPU makes a driving decision which obeys the traffic rules after jointly analyzing the perceived traffic environment information and the AV's driving state.
- **Motion control**: According to the decisions made by the CPU, the LCU generates the corresponding control commands on the accelerator, brake, steering wheel and transmission lever to control the motion of the AV.

Figuratively speaking, the environment perception, behavior decision and motion control can be regarded as the eyes, the brain and the limbs of the AV, respectively. Any errors in the system may make the AV perform wrong operations, resulting in severe losses including money, time and even lives. Therefore, the environment perception part should be as comprehensive as possible to perceive the surrounding environment, that is, the eye must capture accurate road conditions. Behavior decision should depend on an experienced model, namely, the brain must be smart enough to analyze data and make decisions based on the scenes that the eyes see. After the brain makes decisions, the limbs in the system need to do the operations according to the brain's directions in time. Through the efficient cooperation of these three parts, we can enjoy the convenient services provided by AVs on the premise of ensuring safety in the near future.

7.2.2   Autonomous Driving with Vehicular Networks

The vehicular networks, which typically consist of vehicle to vehicle (V2V) and vehicle to infrastructure (V2I) communications, have been studied for several years to improve the efficiency of traffic and provide AVs with entertainment services [32–35]. In the vehicular networks, there are a group of infrastructures deployed along roads and connected with the remote content server to provide content to the AVs within their coverage. In addition, each AV has an on-board unit (OBU) to cache contents based on their interests and can share content with the nearby AVs using V2V communication. Due to the high mobility of AVs and the dynamic change of the network topology, the communication technology used in vehicular networks mainly depends on the dedicated short-range communication i.e., DSRC [36–41].

- **Safety service**: An AV can share the position information with the surrounding AVs to accurately know their statuses and location information to have a safe trip. In addition, infrastructures can broadcast the real-time traffic status of some road segments to the covered AVs to provide a reference for path planning. In this way, the AVs can know the traffic status without the sight and make decisions in advance. As such, timely response can be made to avoid traffic accidents when an emergency occurs.
- **Convenient service**: Each AV can cache its interested content (e.g., the information of nearby gas stations and parking lots) and share the content with other AVs. In addition, the AVs in the coverage of an infrastructure also can obtain content using V2I communication. When an AV needs to request content, if the content is cached in its connected AVs, the AV can obtain this content directly in a single hop. Otherwise, the AV can broadcast its interest and obtain the content in a multi-hop way from an AV/infrastructure that is without the range of the connection. On the other hand, if an AV is within the coverage of an infrastructure, the AV can download its requested content by connecting the infrastructure.

- **Social service**: An AV can share content with others with the same interest to construct the social ties. For example, a group of AVs on the highway can form an ad hoc network for playing online games [42].

Through the cooperation of AVs and infrastructures, not only the safe of autonomous driving can be ensured, but also the pleasure of travel can be increased.

7.2.3 Deep Learning Based Autonomous Driving

To facilitate the autonomous driving, it is necessary to constantly perceive the surrounding environment and obtain a large amount of data for decision making. However, it is difficult to extract feature information from the collected data effectively for modeling the system manually to achieve the desired decision accuracy. The reasons are as follows.

- The surrounding environment changes all the time and the sensors cannot sense all the information around them. For example, when the target is blocked or the weather and light are bad, the sensed data may be incomplete, resulting in a great uncertainty and diversity in the environment perception process.
- The complex scenes sensed by the video camera are hard to define with limited rules and features.
- Even if the feature parameters can be designed manually based on the prior knowledge of designers, it does not make use of the advantages of big data, resulting in low accuracy.

Fortunately, the rapid development of deep learning and its superior performance provide hope for solving this problem [43]. The concept of deep learning derives from the study of artificial neural networks which can be seen as a simulation of the recognition mechanism of human brain. Through the training model to analyze a large number of irregular data sensed from the surrounding environment, deep learning can dig out the intrinsic correlation among data and express it with concise information for decision-making. As shown in Fig. 7.2, deep learning can not only be used to learn and recognize the objects from the data collected from the surrounding environment, but also can be used for the behavior decision and to monitor the status of passengers inside the AV. For example, deep learning can be used to analyze the collected data (e.g., location, distance, etc.) after recognizing objects so as to make more precise behavior decisions to guide the moving of the AV. More importantly, the intelligence of the AV to identify objects and make decisions increases with the increase of the process of data training by using the learning model. For example, TOYOTA recently presented a simple demo system for autonomous driving. There are eight AV models that have no driving experience drive in a square terrain with the barrier and direction. After 4 h of learning, all the AVs can drive safely and almost without accidents. Because of these remarkable features, deep learning is expected to provide safe and convenient driving for AVs.

7.3   Architecture of Deep Learning Based Autonomous Driving in Vehicular Networks

In this section, we propose the architecture of deep learning based autonomous driving in vehicular networks. We first introduce the network architecture, followed by the composition of learning group for collaborative autonomous driving.

7.3.1   Network Architecture

As shown in Fig. 7.3, the architecture of the network consists of the AVs, infrastructures (e.g., RSUs) and remote content server.

- **AVs:** The AVs in the networks have three main functions. (1) Cache and communication: each AV has an OBU through which the AV can cache some contents, such as the high precision map of a part of road segments and some learning algorithms, where the map of each road segment is identified by the number of the road segment and the update time. By using the V2V

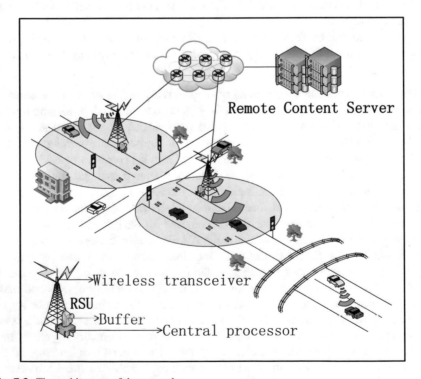

Fig. 7.3 The architecture of the network

communication, AVs can share the contents with each other. (2) Environmental perception: with a group of sensors and video cameras, each AV can perceive the surrounding environment, such as pedestrian, traffic signs, lane markings, etc. [44–48]. (3) Behavior decision and motion control: through the reference of the high precision map, the CPU analyzes the sensed data and outputs the control commands to the control system of the AV. After that, the LCU executes the control commands sent by the CPU.

- **RSUs:** The RSUs are deployed along roads by the network operator and are connected with the remote content server to cache the high precision map, training models and other contents. If a map of a road segment is cached in an RSU, the AV within its coverage can obtain the updated map from the RSU directly. If the map is not available in its cache, the RSU then retrieves the map from the remote content server and then delivers the map to the AV. In addition, the RSUs also cache several training models that are frequently used by AVs. By considering the fact that the cache capacity of each RSU is limited, the maps and training models are cached in the RSU based on the request times.
- **Remote content server:** The remote content server caches the map of all the road segments and can provide different training models for learning. If a map of the road segment is updated, the remote content server then updates the replicas of the map cached in the RSUs. On the other hand, the remote content server can provide AVs with different training models to adapt to the different traffic conditions.

7.3.2   Composition of Learning Group

Based on different situations, the composition of the learning group shows different forms. As shown in Fig. 7.4, the composition of learning groups can be classified by the following cases.

- **Without the help of other AVs/RSUs:** In this case, the AVs in a group drive on road without the help of other AVs or RSUs. As we can see in Fig. 7.4, learning group 1 contains two fixed members to drive on road collaboratively.
- **With the help of other AVs:** In this case, the group needs to obtain a high precision map or training model from other AVs. As such, the group has some fixed members and some temporary members at that time. For example, group 2 in Fig. 7.4 contains two fixed members and one temporary member.
- **With the help of RSUs:** Similar to case 2, the group in this case also has a temporary member for collaborative driving. Unlike case 2, the temporary member in this case is the RSU rather than other AVs. An example of this case is group 3 in Fig. 7.4, where the group is in the coverage of the RSU and the RSU provides a map or training model to the AV.
- **With the help of both AVs and RSUs:** In this case, the group is not in the coverage of any RSUs. However, the group still can ask for help from the nearby

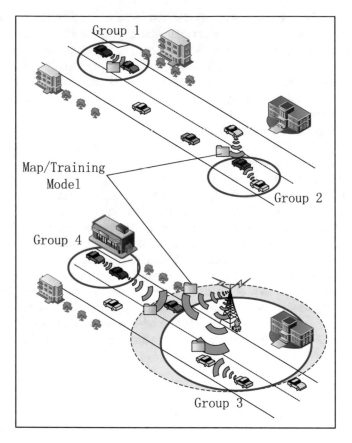

Fig. 7.4 The composition of learning groups

RSU with multi-hop transmission, where the AVs between the group and the RSU are requested for forwarding. As shown in Fig. 7.4, group 4 can obtain its requested map or training model from the RSU with the help of one AV, where both the RSU and the AV are temporary members.

7.4   Learning with Groups: Deep Learning Based Autonomous Driving in Vehicular Networks

In this section, we first introduce the topology of the learning groups based on the proposed architecture. After that, we detail the learning process of groups, followed by the allocation of profits/costs within a group.

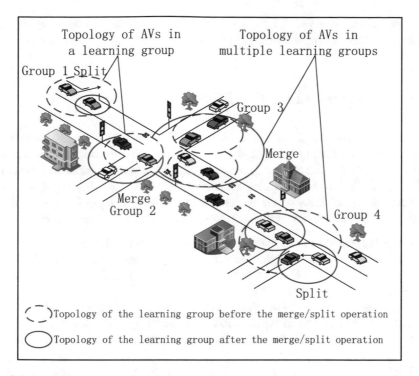

Fig. 7.5 The topology of learning groups

7.4.1   Topology of Learning Groups

In this subsection, we show the topology of learning groups. As the temporary members in the group only provide contents to the group and then leave the group directly, we focus on the fixed members when analyzing the topology of learning groups.

- **Topology of AVs in a learning group:** As shown in Fig. 7.5, if an AV arrives at the destination, the AV can split its current group and drive alone (Group 1). On the contrary, an AV can merge in a group if the AV encounters a learning group that has the same driving direction (Group 2).
- **Topology of AVs in multiple learning groups:** For the AVs in different learning groups, they can merge in a new group for collaborative driving (Group 3). If a number of AVs intend to change their driving direction or arrive at the destination, these AVs then split from their current learning group to form a new one (Group 4).

Note that, before AVs leave their current group or merge in a new group, they need to request the learning group in advance using V2V communication. In this way, as we can see in Fig. 7.5, the topology of AVs in multiple learning groups can be changed flexibly and dynamically by using the merge and split operations.

7.4.2 Cooperative Learning Within a Group

In each learning group, different AVs play different roles in the process of collaborative driving as AVs typically have different learning abilities. The learning ability of AV n is denoted as l_n which follows the uniform distribution within $[l_{min}, l_{max}]$. We define the learning accuracy of AV n as $\log(2 + l_n/L) + \sigma$, where L is an adjustment factor and σ is a random variable and ranges from $[\sigma_{min}, \sigma_{max}]$ to describe the uncertainty in the processes of learning and decision making.

- **Learning process of the leader:** Once a group is formed, the group first elects the leader which has the highest learning ability to improve the driving efficiency. The leader is responsible for the main learning process and needs to perceive the environment ahead of the group.
- **Learning process of the auxiliary AVs:** The auxiliary AVs refer to the AVs that assist the leader to complete the collaborative driving. For example, the AVs on both sides need to provide an auxiliary perception of the environment for the group leader.
- **Learning process of other AVs:** The rest of the AVs in the group only need to coordinate with the group leader and the auxiliary AVs by making decision commands according to their own driving statuses and the decisions made by the leader and auxiliary AVs. Note that, because the hardware systems and locations of these AVs are different, they need to do different operations to adjust their driving statuses to achieve collaborative driving with the leader. For example, there is a pedestrian ahead of the group, the leader decides to decelerate at 5 m/s and broadcast the decision to the members in the group. The members then make deceleration commands according to their current velocity, the distance and angle to the leader, and the performance of their brake systems.
- **Learning with groups:** After the leader completes a learning process (i.e., environment perception, behavior decision and motion control), the leader delivers the data input and decision results to the members in the group. In this way, the leader can teach the members in the group that have low learning ability to drive. With such a learning mechanism, the learning abilities of the members in the group can be improved.

7.4.3 Allocation of Profits/Costs Within a Group

For an AV in a group that drives on the road, it may earn profits by providing relay service or the requested map to other AVs which are without the group. On the other hand, when the group needs the help from other AVs or RSUs, it needs to pay the corresponding costs. For the profit, it is shared among the AVs in the groups based on their different contributions. The cost, however, is equally shared by all the AVs in the group. By considering the issue of ethical responsibility in the process of autonomous driving [49], the AVs which obtain benefits or spend cost in the group need to share responsibility for the accident or other troubles.

- **Profits of an AV in a learning group:** If some AVs in group N provide the relay service to another group M, $M \bigcap N = \phi$, these AVs in group N can obtain the profits paid by group M. Let set N_r be the AVs which provide the relay services to group M. Then the payment that AV $n_r (n_r \in N_r)$ is obtained can be expressed as

$$U_{n_r} = \frac{(k_{in} + 1)(C_r - c_r)}{|N_r|} = C_r - c_r, \tag{7.1}$$

where k_{in} is the hops that the relay service spends within the group. C_r and c_r are the price and cost for one hop, respectively. $|N_r|$ is the number of AVs which provide the relay service. On the other hand, if group N can provide the requested map or training model to another group, the profits of the AVs is based on their contributions to the service. Specifically, the payment obtained by the AV that provides the content to group M can be expressed as

$$\alpha C_{M,V} + \beta C_{T,V} + C_r - c_r, \tag{7.2}$$

where $C_{M,V}$ and $C_{T,V}$ are the prices of the map and training model provided by AVs, respectively. $\alpha = 1$ if the requested content is the high precision map while $\beta = 1$ indicates that the content provided by the AV is the training model. The other AVs, which provide the relay services, can also obtain the payment by $C_r - c_r$.

- **Costs of an AV in a learning group:** In learning group N, different AVs play different roles in the process of environment perception. For the leader in learning group N, its cost can be expressed as

$$B_{N,L} = x \cdot Q \cdot d_N, \tag{7.3}$$

where d_N is the driving distance of group N. x is a random variable. Q is the adjustment factor. For the AVs that used to auxiliary perceive the environment, the cost is

$$B_{N,A} = \sum_{n=1}^{|N'|} y \cdot Q \cdot d_N, \tag{7.4}$$

where y is a random variable. $|N'|$ is the number of auxiliary AVs in the group. For the reminder AVs, the cost is

$$B_{N,R} = \sum_{n=1}^{|N|-|N'|-1} z \cdot Q \cdot d_N, \tag{7.5}$$

where z is a random variable and we have $x > y > z$.

On the other hand, if the group buys the map or training model from other AVs or RSUs, it also spends costs. The costs of this part can be given by

$$B_{N,B} = \mu C_{M,R} + (1 - \mu)C_{M,V} + \varrho C_{T,R} + (1 - \varrho)C_{T,V} + (k_{in}c_r + k_{out}C_r),$$

where k_{out} is the hops that the relay service spends without the group. $\mu = 1$ and $\mu = 0$ indicate that the map is provided by an RSU and an AV, respectively. $C_{M,R}$ and $C_{T,R}$ are the prices of the map and training model declared by RSUs. If the training model is provided by an RSU, we have $\varrho = 1$. By contrast, $\varrho = 0$ indicates that the training model is provided by an AV.

As such, the cost of each AV in the group can be expressed as

$$\frac{B_{N,L} + B_{N,A} + B_{N,R} + B_{N,B}}{|N|}, \tag{7.6}$$

where $|N|$ is the number of AVs in group N. Note that, if the price of the content obtained from one of the AVs in the group is higher than obtaining the content from other sites, the leader obtains the content from other sites to reduce the cost. For example, there are so many AVs in the group and there exists an AV cached the map which is requested by the leader. If $k_{in}c_r > k_{out}C_r + \mu C_{M,R}$, the leader will downloads the map from the RSU rather than from the AV in the group.

7.5 Simulation

In this section, a case is studied to demonstrate the efficiency of our proposal. We first introduce the network setting followed by the analysis of the results.

7.5.1 Setting

Consider a general scenario that an area has 100 road segments and we focus on two schemes according to different autonomous driving situations.

- **Act alone:** In this scheme, an AV drives on a road segment alone, where the road segment is randomly selected.
- **The Proposal:** In our proposal, AVs in the networks learn and drive with groups on the randomly selected road segment.

The uncertainty value in the processes of environment perception and decision making lies in $[-0.1, 0.1]$. The price of the map of a road segment declared by an AV and an RSU are 20 and 50, respectively. The price and cost for relay service are set to be 5 and 2. The values of x, y and z range from $[0.6, 0.9]$, $[0.3, 0.6]$ and $[0.1, 0.3)$, respectively. Under these conditions, the metrics used for the evaluation include:

- **Learning accuracy:** The value of the lowest learning ability of each AV is set to be 10, and we evaluate the average learning accuracy of the AV by changing the highest learning ability of AVs from 10 to 50. The learning ability of each AV is randomly selected between the lowest and the highest learning ability.
- **Cost for sensing surrounding environment:** The number of AVs in the group is increased from 1 to 9 to evaluate the cost of AVs for sensing the surrounding environment.
- **Cost for content sharing:** Each AV caches several maps of the segments in this area and some training models, where the minimum number of maps cached in each AV is 0 and the maximum number changes from 0 to 100 for testing. For simplicity, we assume that each AV uses its own training models for autonomous driving. If an AV or a group does not have the map of the current road segment, the AV or the group needs to request the map from other AVs or an RSU.

7.5.2 Results Analysis

Figure 7.6 shows the average learning accuracy of an AV which acts alone and a group of AVs drive collaboratively. From this figure, we can see that learning with groups can achieve a higher learning accuracy compared with the scheme that an AV acts alone. In addition, the learning accuracy of the group increases with the increase of the number of AVs. This is because with the increase of the AVs' number in the group, the probability that there exists an AV with high learning ability becomes high, resulting in high learning accuracy.

Figure 7.7 shows the cost of each AV for sensing the surrounding environment by changing the number of AVs in the group from 1 to 9. We can see from this figure that our proposal can result in a lower cost for sensing than the scheme that the AV drives by itself. The reasons are as follows. First, AVs play different roles in the group and therefore spend different costs, such as auxiliary AVs' costs for sensing is lower than the leader. Second, the costs for sensing in our proposal are equally shared among the AVs in the group. As we can see, the cost of the group decreases if the number of AVs in the group is increased.

Figure 7.8 is the cost of each AV for obtaining the high precision map. In Fig. 7.8, it can be seen that the proposal can achieve a lower cost than the AV which acts alone. In specific, the more number of AVs in the group, the larger difference between the two schemes as the probability that the map is cached in the group increases with the increase of the number of AVs in the networks. In addition, when

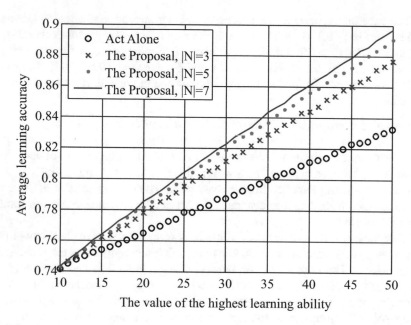

Fig. 7.6 The average learning accuracy by changing the highest learning ability

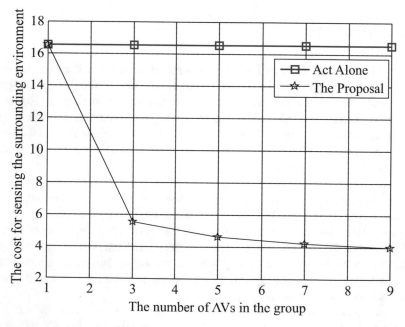

Fig. 7.7 The cost for sensing the environment by changing the number of AVs in the group

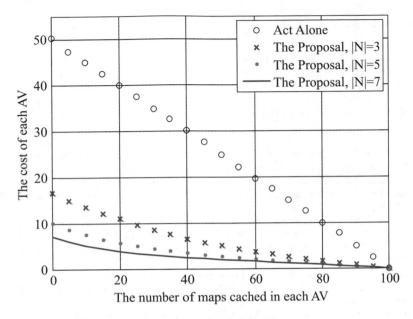

Fig. 7.8 The cost of each AV by changing the number of maps cached in the OBU

the number of maps cached in each AV is 0, the difference between the two schemes is the largest. On the contrary, when each AV in the networks caches all the maps of the road segments, the two schemes have the same cost (i.e., 0) as both of them do not need the help from other AVs or RSUs.

7.6   Summary

In this chapter, we have proposed a collaborative driving scheme in which AVs can learn and drive with groups. To facilitate the autonomous driving, we have presented the architecture of the deep learning based autonomous driving in vehicular networks. Then, we have analyzed the dynamic topology of AVs in a learning group and the topology among multiple learning groups to model the dynamic change of AVs in a group. After this, we have detailed the process of the collaborative driving of AVs with groups, where the leader that has the highest learning ability can guide the AVs to learn together with the target of improving the learning accuracy. Next, we have investigated the profits and costs of groups according to their different contributions to the learning group. Finally, a case has been studied to prove that the proposed deep learning based collaborative autonomous driving framework can efficiently improve the learning accuracy and reduce the costs for both environment perception and content sharing.

References

1. I. Shim, J. Choi, S. Shin, T. Oh, U. Lee, B. Ahn, D. Choi, D.H. Shim, I. Kweon, An autonomous driving system for unknown environments using a unified map. IEEE Trans. Intell. Transp. Syst. **16**(4), 1999–2013 (2015)
2. L. Ma, J. Xue, K. Kawabata, J. Zhu, C. Ma, N. Zheng, Efficient sampling-based motion planning for on-road autonomous driving. IEEE Trans. Intell. Transp. Syst. **16**(4), 1961–1976 (2015)
3. L. Du, Z. Wang, L. Wang, Z. Zhao, F. Su, B. Zhuang, N.V. Boulgouris, Adaptive visual interaction based multi-target future state prediction for autonomous driving vehicles. IEEE Trans. Veh. Technol. **68**(5), 4249–4261 (2019)
4. C. Ma, J. Xue, Y. Liu, J. Yang, Y. Li, N. Zheng, Data-driven state-increment statistical model and its application in autonomous driving. IEEE Trans. Intell. Transp. Syst. **19**(12), 3872–3882 (2018)
5. Q. Luo, Y. Cao, J. Liu, A. Benslimane, Localization and navigation in autonomous driving: threats and countermeasures. IEEE Wirel. Commun. **26**(4), 38–45 (2019)
6. B. Vlasic, U.S. proposes spending $4 billion on self-driving cars. New York Times, Jan. 14 (2016)
7. S. Singh, Critical reasons for crashes investigated in the national motor vehicle crash causation survey, Rep. DOT HS 812 115, National Highway Traffic Safety Administration, Washington, D.C., USA (2014)
8. P. Dai, K. Liu, Q. Zhuge, E.H. Sha, V.C.S. Lee, S.H. Son, Quality-of-experience-oriented autonomous intersection control in vehicular networks. IEEE Trans. Intell. Transp. Syst. **17**(7), 1956–1967 (2016)
9. A. Nanda, D. Puthal, J.J.P.C. Rodrigues, S.A. Kozlov, Internet of autonomous vehicles communications security: overview, issues, and directions. IEEE Wirel. Commun. **26**(4), 60–65 (2019)
10. K. Zheng, Q. Zheng, H. Yang, L. Zhao, L. Hou, P. Chatzimisios, Reliable and efficient autonomous driving: the need for heterogeneous vehicular networks. IEEE Commun. Mag. **53**(12), 72–79 (2015)
11. R.W.L. Coutinho, A. Boukerche, Guidelines for the design of vehicular cloud infrastructures for connected autonomous vehicles. IEEE Wirel. Commun. **26**(4), 6–11 (2019)
12. Z. Su, Y. Hui, Q. Yang, The next generation vehicular networks: a content-centric framework. IEEE Wirel. Commun. **24**(1), 60–66 (2017)
13. C. Shao, S. Leng, Y. Zhang, A. Vinel, M. Jonsson, Performance analysis of connectivity probability and connectivity-aware mac protocol design for platoon-based vanets. IEEE Trans. Veh. Technol. **64**(12), 5596–5609 (2015)
14. G. Xue, H. Zhu, Z. Hu, J. Yu, Y. Zhu, Y. Luo, Pothole in the dark: Perceiving pothole profiles with participatory urban vehicles. IEEE Trans. Mob. Comput. **16**(5), 1408–1419 (2017)
15. C. Wang, J. Wang, Y. Shen, X. Zhang, Autonomous navigation of uavs in large-scale complex environments: a deep reinforcement learning approach. IEEE Trans. Veh. Technol. **68**(3), 2124–2136 (2019)
16. C. Yan, H. Xie, D. Yang, J. Yin, J. Zhang, Q. Dai, Supervised hash coding with deep neural network for environment perception of intelligent vehicles. IEEE Trans. Intell. Transp. Syst. **19**(1), 284–295 (2018)
17. T.Y. He, N. Zhao, H. Yin, Integrated networking, caching and computing for connected vehicles: a deep reinforcement learning approach. IEEE Trans. Veh. Technol. **67**(1), 44–55 (2018)
18. H. Wang, Y. Yu, Y. Cai, X. Chen, L. Chen, Q. Liu, A comparative study of state-of-the-art deep learning algorithms for vehicle detection. IEEE Intell. Transp. Syst. Mag. **11**(2), 82–95 (2019)
19. B. Zhang, C.H. Liu, J. Tang, Z. Xu, J. Ma, W. Wang, Learning-based energy-efficient data collection by unmanned vehicles in smart cities. IEEE Trans. Ind. Inf. **14**(4), 1666–1676 (2018)

20. I. Shim, J. Choi, S. Shin, T. Oh, U. Lee, B. Ahn, D. Choi, D.H. Shim, I. Kweon, An autonomous driving system for unknown environments using a unified map. IEEE Trans. Intell. Transp. Syst. **16**(4), 1999–2013 (2015)

21. L. Ma, J. Xue, K. Kawabata, J. Zhu, C. Ma, N. Zheng, Efficient sampling-based motion planning for on-road autonomous driving. IEEE Trans. Intell. Transp. Syst. **16**(4), 1961–1976 (2015)

22. C. Ma, J. Xue, Y. Liu, J. Yang, Y. Li, N. Zheng, Data-driven state-increment statistical model and its application in autonomous driving. IEEE Trans. Intell. Transp. Syst. **19**(12), 3872–3882 (2018)

23. R.P.D. Vivacqua, M. Bertozzi, P. Cerri, F.N. Martins, R.F. Vassallo, Self-localization based on visual lane marking maps: an accurate low-cost approach for autonomous driving. IEEE Trans. Intell. Transp. Syst. **19**(2), 582–597 (2018)

24. F. Lyu, J. Ren, P. Yang, N. Cheng, W. Tang, Y. Zhang, X.S. Shen, Fine-grained TDMA MAC design toward ultra-reliable broadcast for autonomous driving. IEEE Wirel. Commun. **26**(4), 46–53 (2019)

25. SAE J3016-201401, Taxonomy and definitions for terms related to on-road motor vehicle automated driving systems (2014) [Online]. Available: http://standards.sae.org/j3016_201401/

26. P. Benevides, G. Nico, J. Catalao, P.M.A. Miranda, Bridging insar and gps tomography: a new differential geometrical constraint. IEEE Trans. Geosci. Remote Sens. **54**(2), 697–702 (2016)

27. K.K. So, H. Wong, K.M. Luk, C.H. Chan, Miniaturized circularly polarized patch antenna with low back radiation for GPS satellite communications. IEEE Trans. Antennas Propag. **63**(12), 5934–5938 (2015)

28. O. Kaiwartya, Y. Cao, J. Lloret, S. Kumar, N. Aslam, R. Kharel, A.H. Abdullah, R.R. Shah, Geometry-based localization for GPS outage in vehicular cyber physical systems. IEEE Trans. Veh. Technol. **67**(5), 3800–3812 (2018)

29. C. Sun, H. Zheng, Y. Liu, Analysis and design of a low-cost dual-band compact circularly polarized antenna for GPS application. IEEE Trans. Antennas Propag. **64**(1), 365–370 (2016)

30. S. Manandhar, Y.H. Lee, Y.S. Meng, J.T. Ong, A simplified model for the retrieval of precipitable water vapor from GPS signal. IEEE Trans. Geosci. Remote Sens. **55**(11), 6245–6253 (2017)

31. Y. Jiao, D. Xu, C.L. Rino, Y.T. Morton, C.S. Carrano, A multifrequency GPS signal strong equatorial ionospheric scintillation simulator: Algorithm, performance, and characterization. IEEE Trans. Aerosp. Electron. Syst. **54**(4), 1947–1965 (2018)

32. L. P. Qian, Y. Wu, H. Zhou, X. Shen, Non-orthogonal multiple access vehicular small cell networks: architecture and solution. IEEE Netw. **31**(4), 15–21 (2017)

33. F. Lyu, H. Zhu, H. Zhou, W. Xu, N. Zhang, M. Li, X. Shen, SS-MAC: a novel time slot-sharing MAC for safety messages broadcasting in VANETs. IEEE Trans. Veh. Technol. **67**(4), 3586–3597 (2018)

34. M. Tao, K. Ota, M. Dong, Foud: integrating fog and cloud for 5G-enabled V2G networks. IEEE Netw. **31**(2), 8–13 (2017)

35. Z. Su, Y. Hui, S. Guo, D2D based content delivery with parked vehicles in vehicular social networks. IEEE Wirel. Commun. **23**(4), 90–95 (2016)

36. K. Abboud, H.A. Omar, W. Zhuang, Interworking of DSRC and cellular network technologies for V2X communications: a survey. IEEE Trans. Veh. Technol. **65**(12), 9457–9470 (2016)

37. K.A. Hafeez, A. Anpalagan, L. Zhao, Optimizing the control channel interval of the DSRC for vehicular safety applications. IEEE Trans. Veh. Technol. **65**(5), 3377–3388 (2016)

38. S. Biddlestone, K. Redmill, R. Miucic, U. Ozguner, An integrated 802.11p WAVE DSRC and vehicle traffic simulator with experimentally validated urban (LOS and NLOS) propagation models. IEEE Trans. Intell. Transp. Syst. **13**(4), 1792–1802 (2012)

39. J. Lim, W. Kim, K. Naito, J. Yun, D. Cabric, M. Gerla, Interplay between tvws and dsrc: Optimal strategy for safety message dissemination in vanet. IEEE J. Sel. Areas Commun. **32**(11), 2117–2133 (2014)

40. X. Wu, S. Subramanian, R. Guha, R.G. White, J. Li, K.W. Lu, A. Bucceri, T. Zhang, Vehicular communications using DSRC: challenges, enhancements, and evolution. IEEE J. Sel. Areas Commun. **31**(9), 399–408 (2013)
41. Z. Tong, H. Lu, M. Haenggi, C. Poellabauer, A stochastic geometry approach to the modeling of DSRC for vehicular safety communication. IEEE Trans. Intell. Transp. Syst. **17**(5), 1448–1458 (2016)
42. T.H. Luan, R. Lu, X. Shen, F. Bai, Social on the road: enabling secure and efficient social networking on highways. IEEE Wirel. Commun. **22**(1), 44–51 (2015)
43. J. Wang, H. Zheng, Y. Huang, X. Ding, Vehicle type recognition in surveillance images from labeled web-nature data using deep transfer learning. IEEE Trans. Intell. Transp. Syst. **19**(9), 1–10 (2018)
44. M.A.S. Kamal, T. Hayakawa, J. Imura, Road-speed profile for enhanced perception of traffic conditions in a partially connected vehicle environment. IEEE Trans. Veh. Technol. **67**(8), 6824–6837 (2018)
45. F. Lyu, N. Cheng, H. Zhou, W. Xu, W. Shi, J. Chen, M. Li, DBCC: Leveraging link perception for distributed beacon congestion control in VANETs. IEEE Internet Things J. **5**(6), 4237–4249 (2018)
46. T. Liu, Y. Zhu, R. Jiang, Q. Zhao, Distributed social welfare maximization in urban vehicular participatory sensing systems. IEEE Trans. Mob. Comput. **17**(6), 1314–1325 (2018)
47. J. Guo, B. Song, Y. He, F.R. Yu, M. Sookhak, A survey on compressed sensing in vehicular infotainment systems. IEEE Commun. Surv. Tutorials **19**(4), 2662–2680 (2017)
48. Q. Yuan, H. Zhou, Z. Liu, J. Li, F. Yang, X. Shen, CESense: Cost-effective urban environment sensing in vehicular sensor networks. IEEE Trans. Intell. Transp. Syst. **20**(9), 3235–3246 (2019)
49. J. Borenstein, J. Herkert, K. Miller, Self-driving cars: ethical responsibilities of design engineers. IEEE Technol. Soc. Mag. **36**(2), 67–75 (2017)

Chapter 8
Conclusions and Future Directions

In this chapter, we summarize the book and provide future research directions that related to the next generation vehicular networks.

8.1 Conclusions

In order to pave the way for the next generation vehicular networks, in this book, we have investigated the modeling, algorithm and applications for the integration of the key enabling technologies and vehicular networks. Based on the analysis and discussion provided throughout this book, we then conclude this book as follows.

- Framework of reputation based content delivery in information centric vehicular networks

    Due to vehicles typically have different behaviors in the vehicular networks, an analytical scheme which jointly considers the ICN and the vehicular networks is thus needed to enhance the content delivery performance. Towards this goal, we have proposed a novel framework based on the reputation to facilitate content delivery in information centric vehicular networks. To help both vehicles and infrastructures select the optimal vehicles to join in the content delivery process, we have designed a GUI for each vehicle to manage its interests and reputation value. Then, we have designed the method to calculate the reputation of vehicles and the costs to pay for the content delivery service, respectively. With the incentives, vehicles are encouraged to take part in the content delivery to increase their reputation and earn profits. After this, we have presented the update of vehicles' reputation in information centric vehicular networks and a Bayes based scheme to find the untrustworthy vehicles in the networks. The simulation result has shown the efficiency of the proposed framework with comparisons to the existing scheme.

© Springer Nature Switzerland AG 2021
Z. Su et al., *The Next Generation Vehicular Networks, Modeling, Algorithm, and Applications*, Wireless Networks, https://doi.org/10.1007/978-3-030-56827-6_8

- Framework of contract based edge caching in vehicular networks

    Edge caching has been advocated to deploy edge devices to provide vehicles with content services with low latency time. Due to the massive demands of various kinds of computing services and different social relationships among vehicles, the edge caching based on the social ties becomes important. By analyzing the traffic status and the demands of vehicles, we have proposed a novel contract based framework for edge caching in vehicular networks, where the contracts can be signed between ECDs and social vehicles in advance. Then, we have presented the process of the contract based edge caching in vehicular networks including content caching, content replacement and content delivery among social vehicles. For each ECD, content caching and content replacement are based on three factors which are the contracts signed between vehicles and ECDs, the popularity of contents and the relevance among contents. As for the content delivery among social vehicles, we have considered two different cases according to their social ties. The proposed framework can decrease the transmission latency, enhance the efficiency of networks, and improve the QoE of social vehicles. The simulation result has verified the efficiency of the proposed framework with comparisons to the conventional schemes.

- Framework of Stackelberg game based computation offloading in vehicular networks

    Since the emerging network applications may have different requirements of computing resources, the model of optimal computation offloading becomes important to provide vehicles with the computing services. Specifically, due to the limited computing power of the MEC servers and the need to consume network resources such as power and bandwidth for performing task, each MEC server typically cannot provide offloading services for all the vehicles within its coverage. In addition, for each MEC server in the networks, it needs to study an effective incentive mechanism in order to attract more vehicles to perform task offloading to obtain benefits. To address the above issues, we have proposed a computation offloading scheme based on the Stackelberg game model in vehicular networks. In the scheme, we have established the Stackelberg game model of vehicles and MEC server, where the benefit functions of them are designed. Then, by demonstrating that the vehicle and MEC server have the unique Stackelberg equilibrium solution, we have designed a distributed computation offloading algorithm based on the Stackelberg game which can obtain the optimal game strategies for the vehicle and the MEC server, respectively. Finally, the simulation results have demonstrated the effectiveness and the efficiency of the proposed collaborative computation offloading scheme.

- Framework of auction based secure computation offloading in vehicular networks

    The integration of cloud computing and edge computing has emerged as a new paradigm to provide vehicular task offloading services in vehicular networks. However, a vehicle which intends to complete its computing task typically can offload the task to more than one edge servers. Therefore, how to select the optimal edge server to complete the offloading service is a challenge. In addition, some malicious edge servers may declare unreasonable prices to execute the offloading service. A market mechanism is thus required to constrain the bid prices of edge servers. Besides, the edge servers may maliciously bid with low prices and provide the offloading services with low quality. To address the above mentioned problems, we have proposed an auction based secure task offloading scheme to ensure that the computing tasks requested by vehicles can be safely offloaded and executed. Specifically, to constrain the bids of the edge servers, we have proposed a task offloading scheme based on the first price sealed auction for the edge servers which intend to join in the task offloading process. Then, by considering the service quality of each edge server, we have designed a TSVM based security evaluation and prediction algorithm to evaluate the service quality of edge servers in the cloud layer. The simulation results have shown that the proposed scheme has a higher efficiency than the conventional schemes.

- Framework of bargain game based secure content delivery in vehicular networks

  Vehicular networks have emerged to provide the safety driving for vehicles and improve the driving experience of drivers. With the popularization of the vehicular networks, secure content delivery to protect the vehicles against threats (i.e., outside attacks and inside attacks) becomes a challenge. To this end, we have proposed a security aware content delivery scheme in vehicular networks. Specifically, we have established a trust evaluation scheme for both vehicles and RSUs by introducing AUs to monitor the actions during the process of content delivery. Then, we have proposed a price competitive scheme between vehicles and RSU by using the bargain game to encourage them to improve their trust values and utilities and achieve the secure content delivery. Finally, we have evaluated the proposal using simulation. The result has demonstrated that the proposed scheme can increase the utilities of vehicles and RSUs to deliver content securely compared to the conventional methods.

- Framework of deep learning based autonomous driving in vehicular networks

  Due to the huge data perceived from complicated traffic environment to be analyzed in real-time and the limited computing power of vehicles, it is quite challenging to achieve vehicular networks enabled autonomous driving, which is one of the most attractive applications in the next generation vehicular networks. To this end, we have proposed a deep learning based autonomous driving scheme in which AVs can learn and drive with groups. To facilitate the autonomous driving, we have proposed the architecture of the deep learning based autonomous driving in vehicular networks. Then, we have analyzed the dynamic topology of AVs in a learning group and the topology among multiple learning groups to model the dynamic change of AVs in a group. After this, we have detailed the process of the collaborative driving of AVs with groups, where the leader that has the highest learning ability can guide the AVs learning together with the target of improving the learning accuracy. Next, we have investigated the profits and costs of groups according to their different contributions to the learning group. Finally, a case has been studied to prove that the proposed framework can efficiently improve the learning accuracy and reduce the costs for both environment perception and content sharing.

8.2   Future Research Directions

This book presents the preliminary results on the modeling, algorithm and applications for the integration of vehicular networks and the key enabling technologies. The future research directions are described as follows.

8.2.1   Trading Mechanism in Vehicular Networks

Note that the vehicles are completely independent, it is reasonable to assume that the vehicles are selfish and will not contribute their resources if without sufficient incentives [1–3]. To overcome the selfishness of vehicles, the trading mechanism is needed to encourage them to contribute their resources so that the performance of the vehicular networks can be enhanced. In addition, with the development of autonomous driving [4–6], an electronic trading system to enable vehicles to charge/pay fees according to their own behaviors (e.g., occupy parking space, charging, etc.) is thus necessary to be established. In a brief, the trading mechanism in vehicular networks that can be used for different vehicular scenarios needs to be further studied to facilitate the vehicular applications.

8.2.2   Security and Privacy in Vehicular Networks

In general, vehicle users in the networks have various types and thus have different behaviors which have serious effect on the vehicular service performance. For example, the aggressive vehicle users may refuse to relay or drop the service request and the malicious vehicle users may tamper with the content requested by other vehicles. Additionally, when a vehicle opens the communication interface to connect with other devices, it is easy to be attacked by other aggressive devices [6–8]. As a consequence, vehicles usually have a low will to take part in the vehicular services due to the resources and security considerations [9]. Therefore, by jointly considering the security and privacy of vehicles in vehicular networks, how to provide trusted vehicular service and improve the cooperation among vehicular communication devices to enhance the efficiency of the service system is still an open issue.

8.2.3   Big Data in Vehicular Networks

Caused by the ever-increasing requirements of vehicular applications (e.g., autonomous driving) and mobile services (on-line games), the volume of data which are generated, collected and transmitted by vehicles in vehicular networks has shown the characteristics of big data [10–14]. As analyzed in [15], the data in vehicular networks can well match the 5Vs of big data characteristics which are volume, variety, velocity, value and veracity. By making full use of the big data, vehicular networks can enable variety of advanced applications (e.g., smart city) which can significantly change the mode of the society and the life styles of humans. How to utilize the big data to optimize the network management and satisfy different service requirements therefore becomes an important issue in the next generation vehicular networks.

8.2.4 QoE Aware Services in Vehicular Networks

With the development of autonomous driving and ITS, the future vehicles can be regarded as well-organized robots, where drivers and passengers no longer need to focus on the driving process and thus pay more attention to the service requirements. However, different people typically have different service requirements even for the same trip or the same service [16–18]. Therefore, by analyzing the QoE of different vehicular users, the customized service system to satisfy the needs of different users is required to be investigated. In addition, the QoE of vehicles usually related to the time and the cost spent on the vehicular applications and services. In the vehicular networks, how to balance the time and cost for each vehicle thus needs to be further studied.

8.2.5 Smart Transportation Systems with Vehicular Networks

The rapid development of autonomous driving technology makes it possible for revolutionizing the conventional ITS by constructing the future smart transportation system. Embedded with various advanced sensors, computing units and transmission devices, the controllable AVs can understand the surrounding traffic environment and make accurate driving decisions by themselves [19–21]. By integrating with vehicular networks, the controllable AVs can be collaboratively managed and scheduled with the target of achieving the global scheduling system and satisfying the diverse requirements of drivers. While giving hope to people, the city traffic has shown new features with the integration of vehicular networks and autonomous driving, which are mainly embodied in the massive data collection, caching, transmission and computing. Being faced with these new features, how to integrate the vehicular networks and AVs to establish a comprehensive perception, real-time decision-making and global scheduling system with the target of providing passengers with safe, efficient and comfortable travel experience therefore becomes a future research issue.

8.2.6 Resource Integration and Allocation in Vehicular Networks

In the vehicular networks, a vehicle can request various services from the vehicles within its communication coverage or RSUs. These services typically consume different types of resources, such as caching, computing and transmission. In addition, different nodes including parked vehicles, moving vehicles, RSUs and BSs have different available resources which can be used for supporting vehicular applications. Based on the service requirements and the available resources in the vehicular networks, an integrated resource management architecture needs to be designed. With the integrated resource management architecture, how to dynamically and efficiently allocate the resources with different types by considering the requirements of various vehicular applications therefore becomes an open issue.

References

1. N. Lu, N. Cheng, N. Zhang, X.S. Shen, J.W. Mark, F. Bai, Wi-fi hotspot at signalized intersection: cost-effectiveness for vehicular internet access. IEEE Trans. Veh. Technol. **65**(5), 3506–3518 (2016)
2. C. Lai, K. Zhang, N. Cheng, H. Li, X. Shen, Sirc: a secure incentive scheme for reliable cooperative downloading in highway VANETs. IEEE Trans. Intell. Transp. Syst. **18**(6), 1559–1574 (2017)
3. Z. Su, Q. Xu, Y. Hui, M. Wen, S. Guo, A game theoretic approach to parked vehicle assisted content delivery in vehicular ad hoc networks. IEEE Trans. Veh. Technol. **66**(7), 6461–6474 (2017)
4. D.A. Chekired, M.A. Togou, L. Khoukhi, A. Ksentini, 5G-slicing-enabled scalable SDN core network: toward an ultra-low latency of autonomous driving service. IEEE J. Sel. Areas Commun. **37**(8), 1769–1782 (2019)
5. S. Liu, L. Liu, J. Tang, B. Yu, Y. Wang, W. Shi, Edge computing for autonomous driving: opportunities and challenges. Proc. IEEE **107**(8), 1697–1716 (2019)
6. Z. Su, Y. Hui, T.H. Luan, Distributed task allocation to enable collaborative autonomous driving with network softwarization. IEEE J. Sel. Areas Commun. **36**(10), 2175–2189 (2018)
7. L. Li, J. Liu, L. Cheng, S. Qiu, W. Wang, X. Zhang, Z. Zhang, Creditcoin: a privacy-preserving blockchain-based incentive announcement network for communications of smart vehicles. IEEE Trans. Intell. Transp. Syst. **19**(7), 2204–2220 (2018)
8. X. Liu, Y. Liu, A. Liu, L.T. Yang, Defending on–off attacks using light probing messages in smart sensors for industrial communication systems. IEEE Trans. Ind. Inf. **14**(9), 3801–3811 (2018)
9. L. Yeh, Y. Lin, A proxy-based authentication and billing scheme with incentive-aware multihop forwarding for vehicular networks. IEEE Trans. Intell. Transp. Syst. **15**(4), 1607–1621 (2014)
10. Z. Zhou, H. Yu, C. Xu, Z. Chang, S. Mumtaz, J. Rodriguez, Begin: big data enabled energy-efficient vehicular edge computing. IEEE Commun. Mag. **56**(12), 82–89 (2018)
11. C. Xu, Z. Zhou, Vehicular content delivery: a big data perspective. IEEE Wirel. Commun. **25**(1), 90–97 (2018)
12. K. Lin, J. Luo, L. Hu, M.S. Hossain, A. Ghoneim, Localization based on social big data analysis in the vehicular networks. IEEE Trans. Ind. Inf. **13**(4), 1932–1940 (2017)

13. N. Kumar, S. Misra, J.J.P.C. Rodrigues, M.S. Obaidat, Coalition games for spatio-temporal big data in internet of vehicles environment: a comparative analysis. IEEE Internet Things J. 2(4), 310–320 (2015)
14. S. Garg, A. Singh, K. Kaur, G.S. Aujla, S. Batra, N. Kumar, M.S. Obaidat, Edge computing-based security framework for big data analytics in VANETs. IEEE Netw. 33(2), 72–81 (2019)
15. N. Cheng, F. Lyu, J. Chen, W. Xu, H. Zhou, S. Zhang, X.S. Shen, Big data driven vehicular networks. IEEE Netw. 32(6), 160–167 (2018)
16. C. Xu, F. Zhao, J. Guan, H. Zhang, G. Muntean, QoE-driven user-centric VoD services in urban multihomed p2p-based vehicular networks. IEEE Trans. Veh. Technol. 62(5), 2273–2289 (2013)
17. E. Yaacoub, F. Filali, A. Abu-Dayya, QoE enhancement of SVC video streaming over vehicular networks using cooperative LTE/802.11p communications. IEEE J. Sel. Top. Signal Process. 9(1), 37–49 (2015)
18. M. Zeng, S. Leng, Y. Zhang, J. He, QoE-aware power management in vehicle-to-grid networks: a matching-theoretic approach. IEEE Trans. Smart Grid 9(4), 2468–2477 (2018)
19. R.P.D. Vivacqua, M. Bertozzi, P. Cerri, F.N. Martins, R.F. Vassallo, Self-localization based on visual lane marking maps: an accurate low-cost approach for autonomous driving. IEEE Trans. Intell. Transp. Syst. 19(2), 582–597 (2018)
20. A. Bhat, S. Aoki, R. Rajkumar, Tools and methodologies for autonomous driving systems. Proc. IEEE 106(9), 1700–1716 (2018)
21. S. Kuutti, S. Fallah, K. Katsaros, M. Dianati, F. Mccullough, A. Mouzakitis, A survey of the state-of-the-art localization techniques and their potentials for autonomous vehicle applications. IEEE Internet Things J. 5(2), 829–846 (2018)

Printed in the United States
by Baker & Taylor Publisher Services